U. S. NAVAL ORDNANCE TEST STATION

AN ACTIVITY OF THE BUREAU OF NAVAL WEAPONS

J. I. Hardy, Capt., USN   Wm. B. McLean, Ph.D.
Commander   *Technical Director*

Best Available Copy

BALLISTIC HANDBOOK

by

R. B. Seeley
and
Roy Dale Cole
Aviation Ordnance Department

NOTS TP 3902

China Lake, California
August 1965

NOTS TP 3902

FOREWORD

This report is a compilation of ballistic information on free-fall ordnance in a form easily used by those engaged in fire control system design, weapon design, or weapon system analysis.

The data is presented in the form of approximate solutions to the equations of motion of particle ballistics; numerical data, including ballistic tables, conversion factors, etc.; and nomographs from which many variables of interest can be quickly obtained.

This work was carried out under Bureau of Naval Weapons WepTask R-520-00 001/216-1/F018-02-03, Weapons System Analysis, and WepTask RAVO 8N001 216-1/S171-00-01, Combat Air Weapons System Development, from February 1963 to March 1965.

This report was reviewed for technical accuracy by Dr. Marguerite M. Rogers. Code 3503, U.S. NOTS, invites comments or suggestions concerning this publication.

Released by
N. E. WARD, Head
Aviation Ordnance Department
31 March 1965

Under authority of
WM. B. McLEAN
Technical Director

NOTS Technical Publication 3902

```
Published by..........................................Aviation Ordnance Department
Manuscript...................................................35/MS-216
Collation................. Cover, 71 leaves, 3 dividers, abstract cards
First printing............................................... 225 copies
Security classification..........................................UNCLASSIFIED
```

ABSTRACT

This handbook contains, in a variety of forms, ballistic information on free-fall, unguided weapons of all types, and is intended for use by those engaged in weapon design, fire control system design, and weapon system analysis. The ballistic information is presented in the form of ballistic trajectory equations, tables, graphs, and nomographs, from which many trajecory parameters of interest can be quickly obtained with accuracy sufficient for design and analysis purposes.

ACKNOWLEDGMENT

Acknowledgment is made to Frank Breitenstein, Analysis Branch, Development Division I, for his efforts in supplying the trajectory programs and much of the numerical data presented in this report; and to Hildegard Weinhardt, Analysis Branch, Aircraft Project Division, for her efforts in providing the scale moduli and scale values for individual curves representing the variables from the given original equations, and also for her guidance and assistance in supervising contractor personnel in constructing and plotting the scales presented in this report.

NOTS TP 3902

CONTENTS

I. Introduction... 1
II. Ballistic Equations.. 1
    A. Purpose and Use of Ballistic Sections............................. 1
    B. Coordinate System and Nomenclatures............................... 2
        1. Coordinate System... 2
        2. Nomenclature.. 3
    C. Assumptions... 4
    D. Equations of Motion, Differential Relationships, and Wind
       Effects... 5
        1. Normal Equations of Motion.................................... 5
        2. Differential Relationships.................................... 6
        3. Wind Correction... 7
    E. Solutions to Equations of Motion.................................. 7
        1. Vacuum Case, $D = 0$.. 7
        2. Series Solutions.. 8
        3. Closed Solutions... 13
        4. Partial Derivative Equations................................. 35
III. Numerical Data.. 41
    A. Conversion Factors and Constants................................. 41
    B. Air Data Table... 41
    C. Ballistic Drag Coefficient Functions for Various Bombs........... 44
    D. Ballistic Tables for the Mk 83, Mk 76, and HD-200 Bombs.......... 47
    E. Ballistic Curves and Sensitivity Graphs.......................... 59
    F. Bomb Data.. 85
IV. Nomographs.. 87
    A. General Usage.. 88
    B. Vacuum Solutions... 94
    C. Standard Drag Bombs... 100
    D. Retarded Bombs.. 118

Figures:
1. Coordinate System.. 2
2-17. Ballistic Drag Functions.. 15-30
18. Ballistic Drag Function... 37
19. Ballistic Drag Coefficient Functions................................ 45
20. Ground Range Versus Altitude.. 60
21. Time of Flight Versus Altitude...................................... 61
22. Impact Angle Versus Altitude.. 62
23. Impact Velocity Versus Altitude..................................... 63
24. Lead Angle Versus Altitude.. 64
25. Ground Range Sensitivity to Altitude Change Versus Altitude......... 65
26. Ground Range Sensitivity to Release Angle Change Versus Altitude.... 66
27. Ground Range Sensitivity to Release Velocity Change Versus Altitude... 67

v

28. Ground Range Sensitivity to Drag Function Change Versus Altitude.. 68
29. Time of Flight Sensitivity to Altitude Change Versus Altitude Versus Altitude.. 69
30. Time of Flight Sensitivity to Release Angle Change Versus Altitude.. 70
31. Time of Flight Sensitivity to Release Velocity Change Versus Altitude.. 71
32. Time of Flight Sensitivity to Drag Function Change Versus Altitude.. 72
33. Impact Angle Sensitivity to Altitude Change Versus Altitude.... 73
34. Impact Angle Sensitivity to Release Angle Change Versus Altitude.. 74
35. Impact Angle Sensitivity to Release Velocity Change Versus Altitude.. 75
36. Impact Angle Sensitivity to Drag Function Change Versus Altitude.. 76
37. Impact Velocity Sensitivity to Altitude Change Versus Altitude. 77
38. Impact Velocity Sensitivity to Release Angle Change Versus Altitude.. 78
39. Impact Velocity Sensitivity to Release Velocity Change Versus Altitude.. 79
40. Impact Velocity Sensitivity to Drag Function Change Versus Altitude.. 80
41. Lead Angle Sensitivity to Altitude Change Versus Altitude...... 81
42. Lead Angle Sensitivity to Release Angle Change Versus Altitude. 82
43. Lead Angle Sensitivity to Release Velocity Change Versus Altitude.. 83
44. Sensitivity of Distance Normal to Line-of-Sight to Release Angle Change Versus Altitude.. 84

Tables:
1. Differential Relationships... 6
2. Ballistic Function ψ as $\ln\psi$ Versus $2/3 \rho c K_D X \sec\theta (kX \sec\theta)$. 31
3. Ballistic Function ϕ and $\ln\phi$ Versus $1/2 \rho c K_D X \sec\theta (3/4 kX \sec\theta)$.. 32
4. Ballistic Function ψ_T and $\ln\psi_T$ Versus $\rho c K_D X \sec\theta (3/2 kX \sec\theta)$.. 33
5. Air Data... 42
6. Ballistic Drag Coefficient Functions.. 46
7. Ballistic Data for the Mk 83 Mod 2 and 3/E...................................... 48
8. Ballistic Data for the Mk 76 Mod 0 and 2/N...................................... 52
9. Ballistic Data for the HD-200... 56
10. Bomb Data... 86

I. INTRODUCTION

The ballistic equations and numerical data presented in this handbook are intended for use by those engaged in weapon design, fire control system design, and weapon system analysis. The first section, containing ballistic equations, defines the coordinate system, gives the nomenclature, and states the assumptions leading to the simplified equations of motion of particle ballistics. Approximate solutions to these equations of motion are then given in a variety of forms. The second section concerning numerical data contains some useful conversion factors, ballistic characteristics and drag functions of some current weapons, trajectory tables, and graphs and nomographs from which the desired information may be easily extracted.

II. BALLISTIC EQUATIONS

A. PURPOSE AND USE OF BALLISTIC SECTIONS

This section contains ballistic trajectory equations in a sufficient variety of forms that, given a particular set of independent variables, an unknown dependent variable can be computed. All of the equations, with the exception of the vacuum ballistic equations, are fairly simple approximations, and as such do not yield exact solutions. They could be further refined and expanded to yield greater accuracy over a larger range of independent variables, but only at the expense of rapidly increasing equation complexity. It is hoped a suitable balance between accuracy and complexity has been attained to accomplish the purpose for which this handbook is published.

To use the equations, the ballistic characteristics of the weapon or shape in question must be known or assumed, i.e., its reciprocal ballistic coefficient and its ballistic (or aerodynamic) drag coefficient. In some instances, it is possible to reverse the procedure to determine the approximate ballistic characteristics required of a weapon to cause it to describe a desired trajectory for a given set of release conditions.

The user is again cautioned that the solutions herein are approximate. If accurate solutions are required, numerical integration of the equations of motion on a high-speed digital computer is recommended.

B. COORDINATE SYSTEM AND NOMENCLATURE

1. Coordinate System

The coordinate system used in the ballistic section of this report is shown in Fig. 1. In this system, whose origin (0, 0, 0) is the weapon release point, $\vec{Z}$ lies along the direction of gravity, $\vec{X}$ is horizontal and lies along the direction of the initial horizontal velocity component of the weapon in the *air mass*, and $\vec{Y}$ is normal to the XZ plane, i.e., $\vec{Y} = \vec{Z} \times \vec{X}$, giving a right-handed system. This XYZ system is stationary in the air mass so that if wind exists, the coordinate system is moving with respect to the ground.

It is convenient to define a coordinate system X'Y'Z' which is fixed with respect to the ground, but which initially (at weapon release) coincides with the XYZ system. This fixed system will be useful in accounting for the effects of wind on the weapon trajectory.

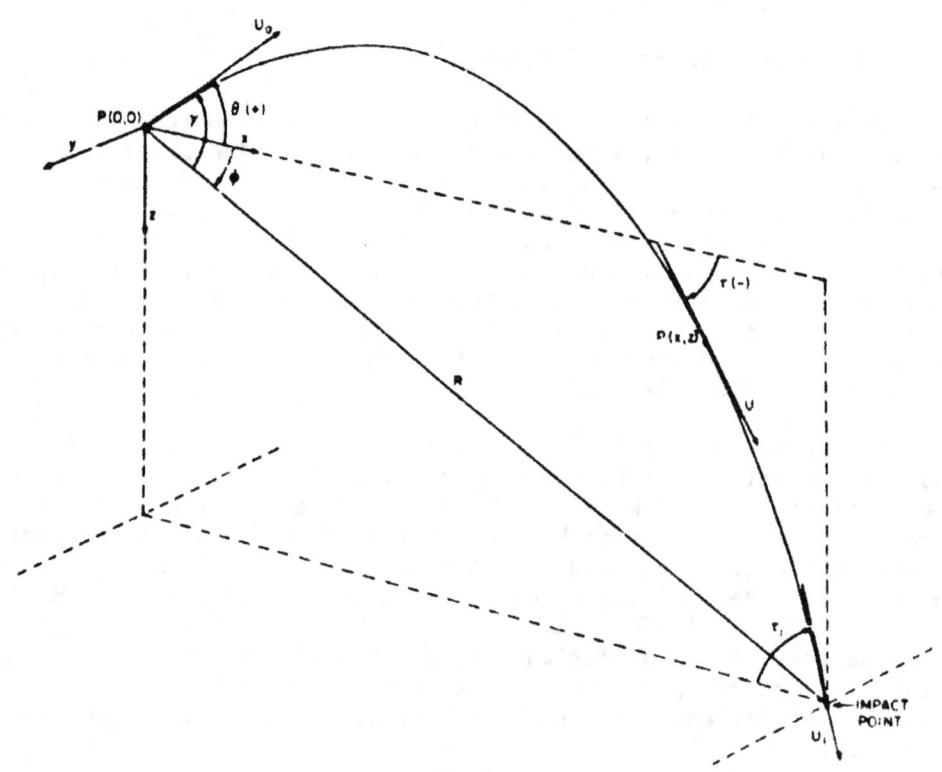

FIG. 1. Coordinate System.

2. Nomenclature

A_1, A_2, A_3	Coefficients of terms in the series solutions of the equations of motion
C_{DO}	Zero lift aerodynamic drag coefficient, a function of Mach number
c	Reciprocal ballistic coefficient equal to id^2/w
c_s	Speed of sound
D	Deceleration of weapon due to aerodynamic drag
d	Maximum body diameter of weapon
g	Acceleration of gravity; 32.174 ft/sec^2
h	Altitude above mean sea level (MSL)
i	Subscript denoting the value of a quantity at weapon impact
i	Form factor of weapon
K_D	Ballistic drag coefficient, a function of Mach number
$K*$	A modified ballistic drag coefficient
o	Subscript denoting an initial condition
p	Subscript denoting a pullup point
R	Slant range to target
t	Time
t_f	Time of flight of weapon
u	Magnitude of weapon speed with respect to the air mass
v	Magnitude of weapon speed with respect to the ground
W	Weapon weight
w	Wind speed with respect to the ground
$\vec{X}, \vec{Y}, \vec{Z}$	Air mass coordinate system; origin at release point. $\vec{Z}$ is positive down, $\vec{X}$ is horizontal, $\vec{Y} = \vec{Z} \times \vec{X}$
$\vec{X}', \vec{Y}', \vec{Z}'$	Coordinate system coinciding with $\vec{X}, \vec{Y}, \vec{Z}$ system of weapon release, but fixed with respect to the ground
α_T	Angle of attack
β_T	Angle of skid
Γ	Function to account for the effects of aerodynamic drag in closed trajectory equations
γ	Ballistic lead angle: angle in a vertical plane between the line of sight to the target at release and the direction of the initial velocity vector

Δ	Operator indicating an incremental change in a variable, i.e., $\Delta X = dX$
θ	Release angle: angle between the velocity vector of the weapon at release and the horizontal, positive when the velocity vector is above the horizontal
ρ	Air density
τ	Angle between the tangent to the trajectory at any point and the horizontal: $\tau_o = \theta$; $\tau_i =$ impact angle
φ	Line of sight or harp angle: angle between the horizontal and the line of sight to the target at release
ψ, Φ, ψ_τ	Functions to account for the effects of aerodynamic drag in closed trajectory equations

C. ASSUMPTIONS

The ballistic equations of the next sections were derived using the following assumptions:

a. The acceleration of gravity is equal to 32.174 ft/sec² and is constant, independent of altitude.

b. The earth is flat and nonrotating.

c. Wind is constant from release point to impact point.

d. The air density is that given by the ICAO Tables of 1954.

e. The forces acting on the weapon are due only to gravity and to the motion of the weapon through the atmosphere.

The aerodynamic forces are further assumed to act only along the longitudinal axis of the weapon with a magnitude given by:

$$F = \frac{1}{2} \rho u^2 S C_{D_o} \quad . \tag{1}$$

In eq. 1, C_{D_o} is the aerodynamic drag coefficient at zero lift or zero angle of attack, and S is a reference area, usually the maximum body cross-section area. The deceleration of the weapon is

$$D = \frac{1}{2} \rho \frac{u^2 S}{W} C_D \tag{2}$$

where W is the weight of the weapon, and the units of ρ have changed from mass per volume in eq. 1 to weight per volume in eq. 2 to account for the "missing" gravity term.

If $S = \frac{\pi d^2}{4}$;

then
$$D = \frac{\pi}{8} \rho u^2 \frac{d^2}{W} C_D . \tag{3}$$

In ballistic work, the weapon deceleration due to drag is normally written

$$D = \rho c u^2 K_D \tag{4}$$

where c, the reciprocal ballistic coefficient, is defined by:

$$c = i \frac{d^2}{W} \tag{5}$$

and K_D is the ballistic drag coefficient. From eq. 3, 4, and 5, it is evident that, for $i = 1$,

$$K_D = \frac{\pi}{8} C_D = 0.3927 \, C_D$$

or

$$C_D = 2.546 \, K_D . \tag{6}$$

In eq. 5, i is called the form factor, and is introduced as a correction term relating the ballistic drag coefficient of some standard projectile to that of a projectile of usually similar shape. Thus, if K_{D_s} is the drag function of a standard projectile, the form factor i is given by

$$i = \frac{K_D}{K_{D_s}}$$

and is a sort of average correction term applied to the drag coefficient of the standard projectile to obtain a drag function suitable for the weapon at hand.

D. EQUATIONS OF MOTION, DIFFERENTIAL RELATIONSHIPS, AND WIND EFFECTS

1. <u>Normal Equations of Motion</u>

In terms of the air mass coordinate system defined, the assumptions stated in the previous section lead to the following differential equations of motion for particle ballistics:

$$\frac{d^2 X}{dt^2} = -D \cos \tau , \qquad \frac{d^2 Z}{dt^2} = D \sin \tau + g ,$$

$$\frac{dX}{dt} = u \cos \tau, \qquad \frac{dZ}{dt} = -u \sin \tau, \qquad (7)$$

where

$$D = \rho c K_D u^2 .$$

The ballistic equations of the next sections were obtained from Taylor series solutions of eq. 7 of the general form

$$\S = \S_0 + \left(\frac{d\S}{d\eta}\right)_0 \eta + \left(\frac{d^2\S}{d\eta^2}\right)_0 \frac{\eta^2}{2!} + \ldots + \left(\frac{d^n\S}{d\eta^n}\right)_0 \frac{\eta^n}{n!} +$$

$$\frac{1}{n!} \int_0^\eta (\eta - \epsilon)^n \frac{d\S^{(n+1)}}{d\epsilon^{(n+1)}} d\epsilon . \qquad (8)$$

where the subscript o indicates that the derivatives are evaluated at the release point of the weapon, i.e., at $X = Y = Z = t = 0$.

2. Differential Relationships

Table 1 is a compilation of some useful first derivatives from which the higher order derivatives can be obtained.

TABLE 1. Differential Relationships

$\frac{d \rightarrow}{d \downarrow}$	X	Z	u	τ	t
X	1	$-\tan \tau$	$-\left(\frac{D + g \sin \tau}{u \cos \tau}\right)$	$\frac{-g}{u^2}$	$\frac{1}{u \cos \tau}$
Z	$-\cot \tau$	1	$\frac{D + g \sin \tau}{u \sin \tau}$	$\frac{g \cot \tau}{u^2}$	$\frac{-1}{u \sin \tau}$
u	$\frac{-u \cos \tau}{D + g \sin \tau}$	$\frac{u \sin \tau}{D + g \sin \tau}$	1	$\frac{1}{u}\left(\frac{g \cos \tau}{D + g \sin \tau}\right)$	$\frac{-1}{D + g \sin \tau}$
τ	$-\frac{u^2}{g}$	$\frac{u^2 \tan \tau}{g}$	$u\left(\frac{D + g \sin \tau}{g \cos \tau}\right)$	1	$\frac{-u}{g \cos \tau}$
t	$u \cos \tau$	$-u \sin \tau$	$-(D + g \sin \tau)$	$-\frac{g \cos \tau}{u}$	1

$$D = \rho c K_D(M) u^2, \qquad \frac{d^2 X}{dt^2} = -D \cos \tau, \qquad \frac{dX}{dt} = u \cos \tau,$$
$$\frac{d^2 Z}{dt^2} = D \sin \tau + g, \qquad \frac{dZ}{dt} = -u \sin \tau.$$

3. Wind Correction

With the normal equations of particle ballistics solved in an air-mass coordinate system, the motion of the air mass must be accounted for if the variables of the trajectory need to be described in terms of a fixed coordinate system, e.g., the X'Y'Z' system defined in Section B.1.

If wind is restricted to be horizontal and constant, the relations between the trajectory parameters in the X'Y'Z' and the XYZ systems are:

$$t_f' = t_f \quad \text{(no vertical wind)}, \tag{9}$$

$$X' = X + W_x t_f, \tag{10}$$

$$Y' = W_y t_f, \tag{11}$$

$$v_i = (u_i^2 + W^2 + 2Wu_i \cos \tau_i)^{1/2}, \tag{12}$$

$$\sin \tau_i' = \frac{u_i}{v_i} \sin \tau_i. \tag{13}$$

In the above equations, W_x and W_y are the X and Y components of wind, respectively, and thus

$$V_x = U_x + W_x,$$
$$V_y = W_y,$$
and
$$V_z = U_z. \tag{14}$$

Also,
$$V_i = (V_x^2 + V_y^2 + V_z^2)^{1/2},$$
$$U_i = (U_x^2 + U_z^2)^{1/2},$$
$$W = (W_x^2 + W_y^2)^{1/2}. \tag{15}$$

E. SOLUTIONS TO EQUATIONS OF MOTION

1. Vacuum Case, D = 0

a. Variable Z

$$X = \frac{U_o^2 \cos \theta}{g} \left[\sin \theta + \sqrt{\sin^2 \theta + \frac{2Zg}{U_o^2}} \right]. \tag{16}$$

$$t = \frac{U_o}{g} \left[\sin \theta + \sqrt{\sin^2 \theta + \frac{2Zg}{U_o^2}} \right]. \tag{17}$$

$$\tan \tau = -\sec \theta \sqrt{\sin^2 \theta + \frac{2Zg}{U_o^2}}. \tag{18}$$

$$U = \sqrt{U_o^2 + 2Zg} = U_o \sqrt{1 + \frac{2Zg}{U_o^2}}. \tag{19}$$

b. Variable t

$$X = U_0 t \cos \theta \ . \tag{20}$$

$$Z = -U_0 t \sin \theta + \frac{gt^2}{2} \ . \tag{21}$$

$$\tan \tau = -\tan \theta + \frac{gt}{U_0 \cos \theta} \ . \tag{22}$$

$$U = \sqrt{U_0^2 + gt(gt - 2U_0 \sin \theta)} \ . \tag{23}$$

c. Variable X

$$Z = -\tan \theta + \frac{gX^2}{2U_0^2 \cos^2 \theta} \ . \tag{24}$$

$$t = \frac{X}{U_0 \cos \theta} \ . \tag{25}$$

$$\tan \tau = -\tan \theta + \frac{gX}{U_0^2 \cos^2 \theta} \ . \tag{26}$$

$$U = \sqrt{U_0^2 + \frac{gX}{U_0 \cos \theta}\left(\frac{gX}{U_0 \cos \theta} - 2U_0 \sin \theta\right)} \ . \tag{27}$$

2. Series Solutions

a. Coefficients of Series Solutions

The series solutions in the next sections are written in terms of coefficients A_1, A_2, and A_3. These coefficients, given below, were evaluated assuming that (1) the drag coefficient K_D remains at its value at release, and (2) air density at altitude h above MSL is given by[1]

$$\rho = \rho_0 e^{-ah} \tag{28}$$

where, for h in feet, $a = 3.015 \times 10^{-5}$ ft^{-1}, and ρ_0 is air density at sea level.

With $D_0 = \rho c K_D U_0^2$, the coefficients are as follows:

$$A_1 = \frac{4 D_0}{U_0^2 \cos \theta} \ , \tag{29}$$

---

[1] This is not in accordance with ICAO standards, but it allows an analytical expression for the density in the series coefficients. This expression gives values very close to the ICAO that was used in computation of the tables.

$$A_2 = \frac{4 D_o}{U_o^4 \cos^2 \theta} \left[2 D_o - (g + a U_o^2) \sin \theta \right], \quad (30)$$

and

$$A_3 = \frac{4 D_o}{U_o^6 \cos^3 \theta} \Big[4 D_o^2 + g (g + a U_o^2) \cos^2 \theta$$

$$- 2 D (4g + 3 a U_o^2) \sin \theta + 2a_o^2 (3g + 2a_o^2) \sin^2 \theta \Big]. \quad (31)$$

Alternate expressions for these coefficients are:

$$A_1 = 4 \rho c K_D \sec \theta, \quad (32)$$

$$A_2 = \frac{A_1^2}{2} - A_1 (g/U_o^2 + a) \tan \theta, \quad (33)$$

$$A_3 = \frac{A_1^3}{4} - \frac{A_1^2}{2} \left(\frac{4g}{U_o^2} + 3a \right) \tan \theta +$$

$$A_1 \left[\frac{g}{U_o^2} \left(\frac{g}{U_o^2} + a \right) + a \left(\frac{3g}{U_o^2} + a \right) \tan^2 \theta \right]. \quad (34)$$

b. **Independent Variable X, Horizontal Range**

$$Z = -X \tan \theta + \frac{gX^2}{2U_o^2 \cos^2 \theta} \left(1 + \frac{A_1 X}{3!} + \frac{A_2 X^2}{4!} + \frac{A_3 X^3}{5!} + \ldots \right). \quad (35)$$

$$t = \frac{X}{U_o \cos \theta} \Bigg[1 + \frac{2 A_1 X}{(2)(8)} + \frac{1}{(3)(8)} \left(A_2 - \frac{A_1^2}{4} \right) X^2 +$$

$$\frac{1}{(4)(8)} \left(\frac{A_3}{3} - \frac{A_1 A_2}{4} + \frac{A_1^3}{16} \right) X^3 + \ldots \Bigg]. \quad (36)$$

$$\tan \tau = \tan \theta - \frac{gX}{U_o^2 \cos^2 \theta} \left(1 + \frac{A_1}{2} \frac{X}{2!} + \frac{A_2}{2} \frac{X^2}{3!} + \frac{A_3}{2} \frac{X^3}{4!} + \ldots \right) \quad (37)$$

$$U \cos \tau = U_o \cos \theta \Bigg[1 - \frac{2A_1 X}{8} - \frac{1}{8} \left(A_2 - \frac{3A_1^2}{4} \right) X^2 -$$

$$\frac{1}{8} \left(\frac{A_3}{4} - \frac{3A_1 A_2}{4} + \frac{5A_1^3}{16} \right) X^3 + \ldots \Bigg]. \quad (38)$$

c. **Independent Variable t, Time of Flight**

$$X = U_o t \cos\theta \left(1 - \frac{A_1 U_o t \cos\theta}{8} - \frac{A_2 - A_1^2}{3 \cdot 8} U_o^2 t^2 \cos^2\theta + \ldots \right). \quad (39)$$

$$\frac{Z}{X} = -\tan\theta + \frac{gt}{2U_o \cos\theta} \left(1 + \frac{A_1 U_o t \cos\theta}{4!} + 0 + \ldots \right). \quad (40)$$

$$\tan\tau = \tan\theta - \frac{gt}{U_o \cos\theta} \left[1 + \frac{A_1 U_o t \cos\theta}{8} + \right.$$
$$\left. \frac{1}{3 \cdot 8}\left(A_2 - \frac{A_1^2}{2}\right) U_o^2 t^2 \cos^2\theta + \ldots \right]. \quad (41)$$

$$U \cos\tau = U_o \cos\theta \left(1 - \frac{A_1 U_o t \cos\theta}{4} - \frac{A_2 - A_1^2}{8} U_o^2 t^2 \cos^2\theta + \ldots \right). \quad (42)$$

d. **Independent Variable $(\tan\theta + Z/X)$**

$$X = \frac{2U_o^2 \cos^2\theta}{g} \left(\tan\theta + \frac{Z}{X}\right) \left[1 - \frac{2A_1}{3!} \frac{U_o^2 \cos^2\theta}{g}\left(\tan\theta + \frac{Z}{X}\right) - \right.$$
$$\left. \left(A_3 - \frac{2A_1^2}{3}\right)\left(\frac{U_o^2 \cos^2\theta}{g}\right)^2 \left(\tan\theta + \frac{Z}{X}\right)^2 + \ldots \right]. \quad (43)$$

$$t = \frac{2U_o \cos\theta}{g} \left(\tan\theta + \frac{Z}{X}\right) \left[1 - \frac{2A_1}{4!}\left(\frac{U_o^2 \cos^2\theta}{g}\right)\left(\tan\theta + \frac{Z}{X}\right) + \right.$$
$$\left. \frac{8}{4!} A_1^2 \left(\frac{U_o^2 \cos^2\theta}{g}\right)^2 \left(\tan\theta + \frac{Z}{X}\right) + \ldots \right]. \quad (44)$$

$$\tan\tau = \tan\theta - 2\left(\tan\theta + \frac{Z}{X}\right)\left[1 + \frac{4A_1}{4!}\left(\frac{U_o^2 \cos^2\theta}{g}\right)\left(\tan\theta + \frac{Z}{X}\right) + \right.$$
$$\left. \frac{4}{4!}\left(A_2 - \frac{2A_1^2}{3}\right)\left(\frac{U_o^2 \cos^2\theta}{g}\right)^2 \left(\tan\theta + \frac{Z}{X}\right) + \ldots \right]. \quad (45)$$

$$U \cos\tau = U_o \cos\theta \left[1 - \frac{2A_1}{4}\left(\frac{U_o^2 \cos^2\theta}{g}\right)\left(\tan\theta + \frac{Z}{X}\right) - \right.$$
$$\left. \left(\frac{A_2}{2} - \frac{A_1^2}{4!}\right)\left(\frac{U_o^2 \cos^2\theta}{g}\right)^2 \left(\tan\theta + \frac{Z}{X}\right)^2 + \ldots \right]. \quad (46)$$

e. <u>Independent Variable ($\tan \theta - \tan \tau$)</u>

$$X = \frac{U_0^2 \cos^2 \theta}{g} (\tan \theta - \tan \tau) \left[1 - \frac{A_1}{4} \left(\frac{U_0^2 \cos^2 \theta}{g} \right) (\tan \theta - \tan \tau) - \frac{1}{4} \left(\frac{A_2}{3} - \frac{A_1^2}{2} \right) \frac{U_0^4 \cos^4 \theta}{g^2} (\tan \theta - \tan \tau)^2 + \ldots \right]. \quad (47)$$

$$\tan \theta - \frac{Z}{X} = \frac{(\tan \theta - \tan \tau)}{2} \left[1 - \frac{2 A_1}{4!} \left(\frac{U_0^2 \cos^2 \theta}{g} \right) (\tan \theta - \tan \tau) - \left(\frac{(A_2 - A_1^2)}{4!} \right) \left(\frac{U_0^4 \cos^4 \theta}{g^2} \right) (\tan \theta - \tan \tau)^2 + \ldots \right]. \quad (48)$$

$$t = \frac{U_0 \cos \theta}{g} (\tan \theta - \tan \tau) \left[1 - \frac{A_1}{8} \left(\frac{U_0^2 \cos^2 \theta}{g} \right) (\tan \theta - \tan \tau) - \frac{1}{4} \left(A_2 - \frac{5 A_1^2}{4} \right) \left(\frac{U_0^2 \cos^2 \theta}{g} \right) (\tan \theta - \tan \tau)^2 + \ldots \right]. \quad (49)$$

$$U \cos \tau = U_0 \cos \theta \left[1 - \frac{A_1}{4} \left(\frac{U_0^2 \cos^2 \theta}{g} \right) (\tan \theta - \tan \tau) - \frac{1}{8} \left(A_2 - \frac{5 A_1^2}{4} \right) \left(\frac{U_0^2 \cos^2 \theta}{g} \right) (\tan \theta - \tan \tau)^2 + \ldots \right]. \quad (50)$$

f. <u>Accuracy of Series Solutions</u>

In general, these series solutions will provide solutions accurate to within one or two percent of value for low or medium drag weapons (e.g. the Mk 76 bomb) released at less than 1,000 ft/sec velocity below +10 degrees release angles, and for ground ranges less than 15,000 to 20,000 feet.

For higher drag weapons, the series convergence is slow, and the remainder term implicit in each of the equations becomes large, causing considerable error. A sample calculation using set E.2.b. will illustrate.

From Table 8, we find for the Mk 76 bomb that the release conditions:

$Z = 5,000$ ft , $U_0 = 800$ ft/sec , $\theta = -20$ deg

yield the resulting trajectory parameters at $Z = 0$:

NOTS TP 3902

$$X = 7{,}986 \text{ ft}, \quad U_i = 818.0 \text{ ft/sec},$$
$$t_f = 11.841 \text{ sec}, \quad \tau = 42.96 \text{ deg}.$$

With the given ground range X and using eq. 24 to get an approximate value of Z from which to find air density, Mach number, and the appropriate value of K_D, the series solutions, set E.2.b., give from eq. 35, 36, 37, and 38:

$$Z = 4{,}992 \text{ ft; error} = 0.16\%, \quad \tau = 42.92 \text{ deg; error} = 0.09\%,$$
$$t_f = 11.803 \text{ sec; error} = 0.32\%, \quad U_i = 824.9 \text{ ft/sec; error} = 0.84\%$$

By contrast, the vacuum equations yield:

$$Z = 4{,}722 \text{ ft; error} = 5.56\%, \quad \tau = -39.30; \text{ error} = 8.69\%,$$
$$t_f = 10.623 \text{ sec; error} = 10.30\%, \quad U_i = 971.5 \text{ ft/sec; error} = 18.77\%$$

As an example for a high drag weapon, Table 9 gives for the fictitious HD 200 bomb, with the release conditions:

$$Z = 500 \text{ ft},$$
$$\theta = 0 \text{ deg},$$
$$U_o = 800 \text{ ft/sec},$$

the following trajectory parameters at Z = 0 ft:

$$X = 2{,}100 \text{ ft}, \quad \tau = 43.75 \text{ deg},$$
$$t_f = 7.053 \text{ sec}, \quad U_i = 184.0 \text{ ft/sec}.$$

Equations set E.2.b. with sea level air density assumed, gives

$$Z = 401.6 \text{ ft; error} = 19.7\%, \quad \tau_i = 33.82 \text{ deg; error} = 22.7\%,$$
$$t_f = 6.496 \text{ sec; error} = 8.2\%, \quad U_i = \text{(unknown, series alternating)}$$

In this example, it is obvious that the convergence of these series solutions is slow, and too few terms are included.

For high drag weapons such as the HD 200, it is suggested that the closed solution equations of the following sections be used rather than the series solutions.

3. Closed Solutions

A closed solution is easier to work with, particularly when the equations are used in a bombing system. Two forms are considered here: (1) an exponential approximation to the series solutions; and (2) an empirical function representation of the solution.

The basic equations are written as:

$$Z = -X \tan \theta + \frac{gX^2}{2U_o^2 \cos^2 \theta} \psi , \qquad (51a)$$

$$t_f = \frac{X}{U_o \cos \theta} \phi , \qquad (51b)$$

$$\tan \tau = \tan \theta - \frac{gX}{U_o^2 \cos^2 \theta} \psi_\tau , \qquad (51c)$$

and

$$U \cos \tau = U_o \cos \theta \, \psi_u . \qquad (51d)$$

The problem is to find convenient functions for ψ, ϕ, ψ_τ, and ψ_u. These are given in series form by eq. 35 through 37. These series suggest that the variable:

$$\rho c K_D X \sec \theta$$

be used in finding suitable functions. For condensing the notation use:

$$k = (2/3) \rho c K_D , \qquad (52)$$

$$k_o = (2/3) \rho_o c K_D , \qquad (53a)$$

and

$$k_T = (2/3) \rho_T c K_D , \qquad (53b)$$

when the T subscript indicates value at the impact (target) point. Figure 2 indicates the bomb terminal velocity as a function of k for ρ equal to sea level value.

One set of approximations is:

$$\ln \psi \doteq kX \sec \theta (1 + 0.285 \, kX \sec \theta) , \qquad (54a)$$

$$\ln \phi \doteq (3/4) kX \sec \theta (1.0364 + 0.134 \, kX \sec \theta) , \qquad (54b)$$

$$\ln \psi_\tau \doteq (3/2) kX \sec \theta (1.0173 + 0.296 \, kX \sec \theta) . \qquad (54c)$$

For many uses these can be approximated by[2]:

$$\ln \psi = kX \sec \theta , \qquad (55a)$$

$$\ln \phi = (3/4)\ln \psi , \qquad (55b)$$

$$\ln \psi_\tau = (3/2)\ln \psi , \text{ and} \qquad (55c)$$

$$\ln \psi_u = (3/2)\ln \psi . \qquad (55d)$$

For other uses, such as in bombing systems and some nomographs, a mathematical expression is not needed. This leads to the empirical functions[3]:

$$\psi = \psi(kX \sec \theta) \qquad (56)$$

$$\phi = \phi(kX \sec \theta) \qquad (57)$$

$$\psi_\tau = \psi_\tau(kX \sec \theta) \qquad (58)$$

Figures 3 to 8 show empirical functions which were determined by using bombing tables (for several bombs) that had been computed by numerical integration, and then computing the functions, such as ψ, in reverse. Figures 9 to 17 show the data used in obtaining the functions. The accuracy of the fit can be seen. Tables 2, 3, and 4 give the functions in number form. Using k_T in the above equations gives a better fit for retarded bombs.

Most retarded bombs are <u>not</u> retarded from release (as is assumed in this report) but are low drag for a short time or distance from the aircraft. This discontinuous drag presents a problem. Empirical functions similar to those shown in Fig. 3-8 can be obtained; however, each bomb will probably have its own set of functions.

Often the lead angle, γ, is desired. This can be obtained by:

$$\sin \gamma = \frac{gX \cos (\gamma - \theta)}{2U_o^2 \cos \theta} \psi . \qquad (59)$$

---

[2] These give good results for low drag bombs. For higher drag bombs and retarded weapons an empirical value of k should be used. The expression for u must be used with skepticism on high drag bombs.

[3] The basic equations 51 and 59 can be used to express the empirical functions independent variable in various forms.

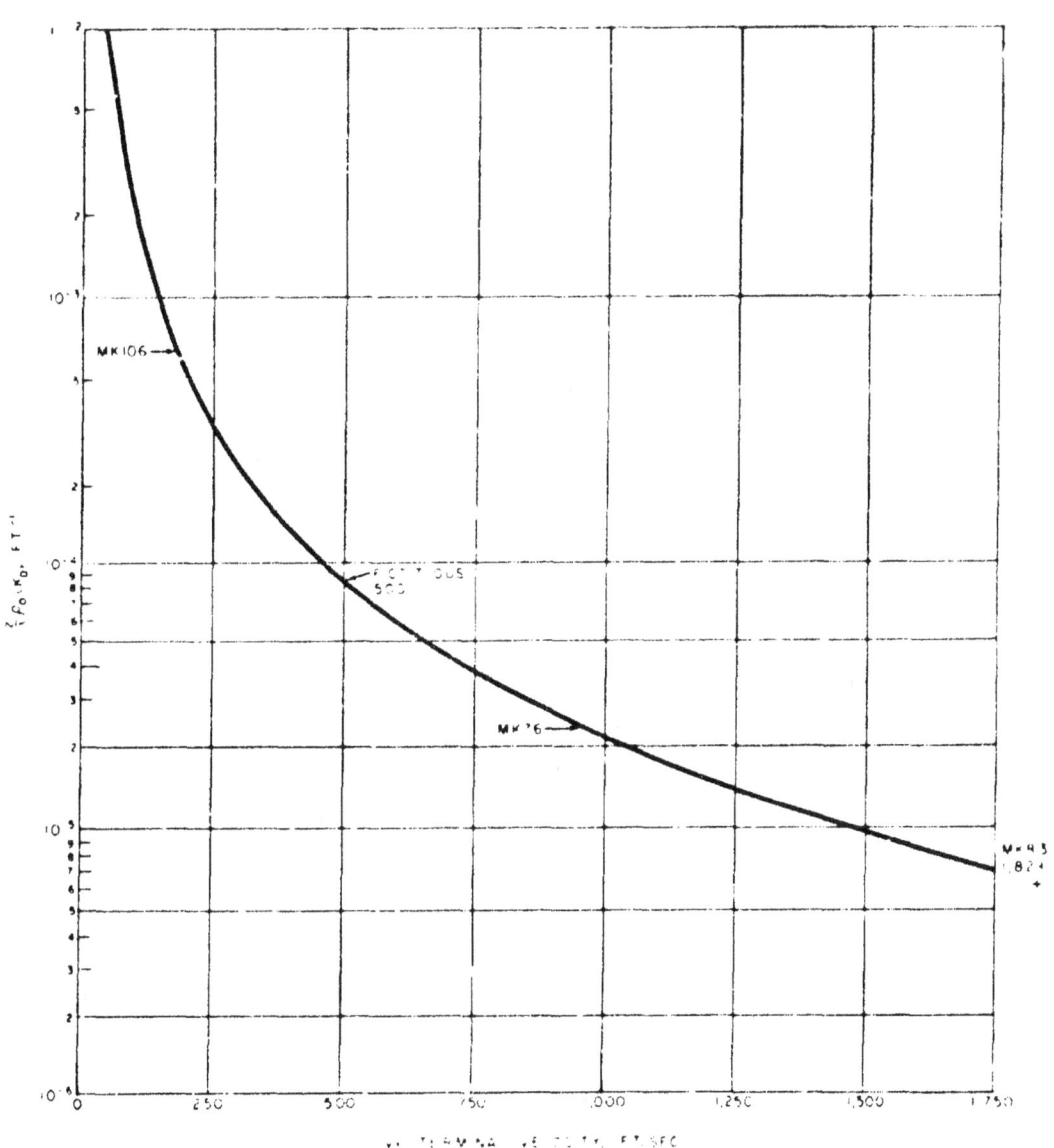

Fig. 2. Ballistic Drag Function.

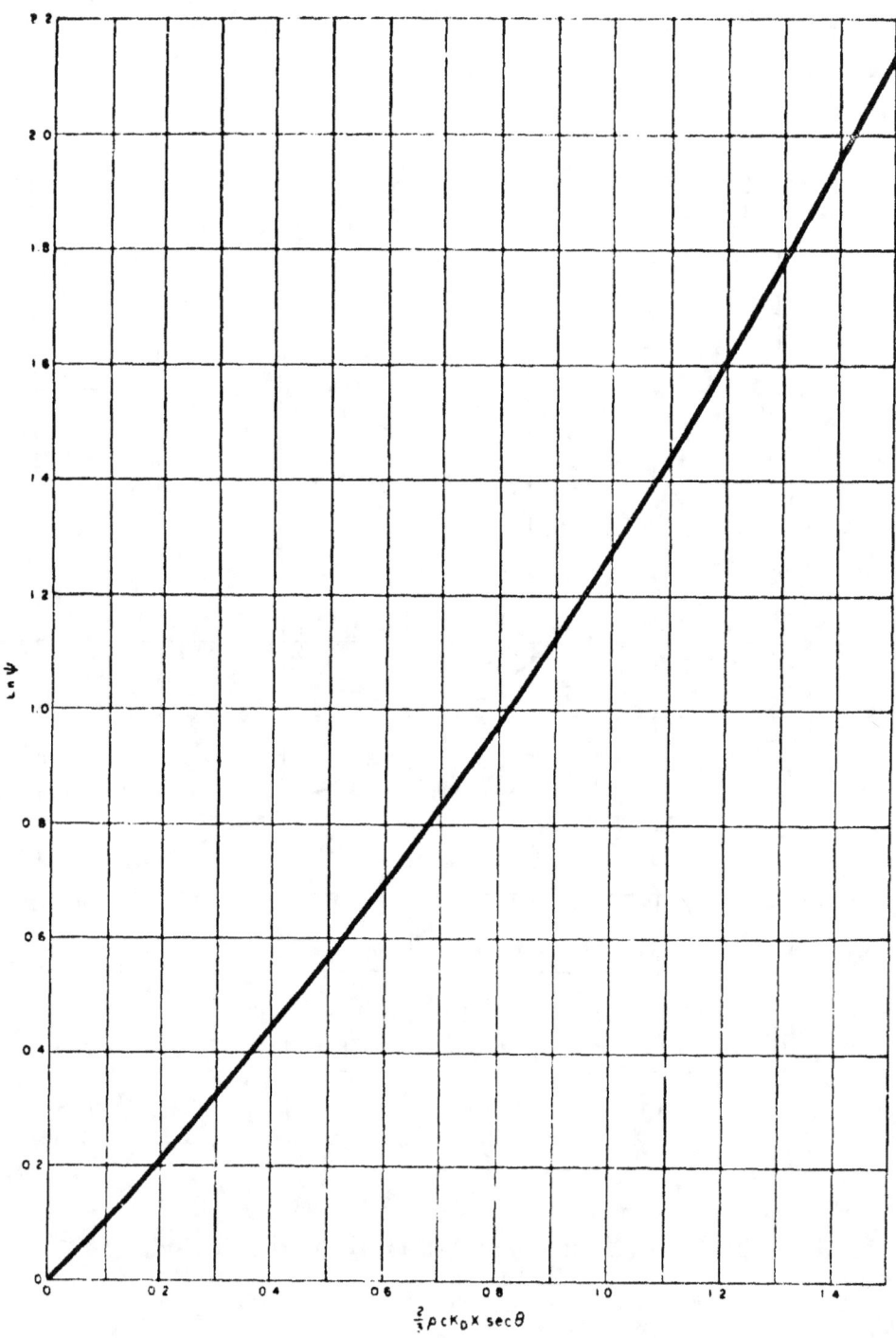

FIG. 3. Ballistic Drag Function.

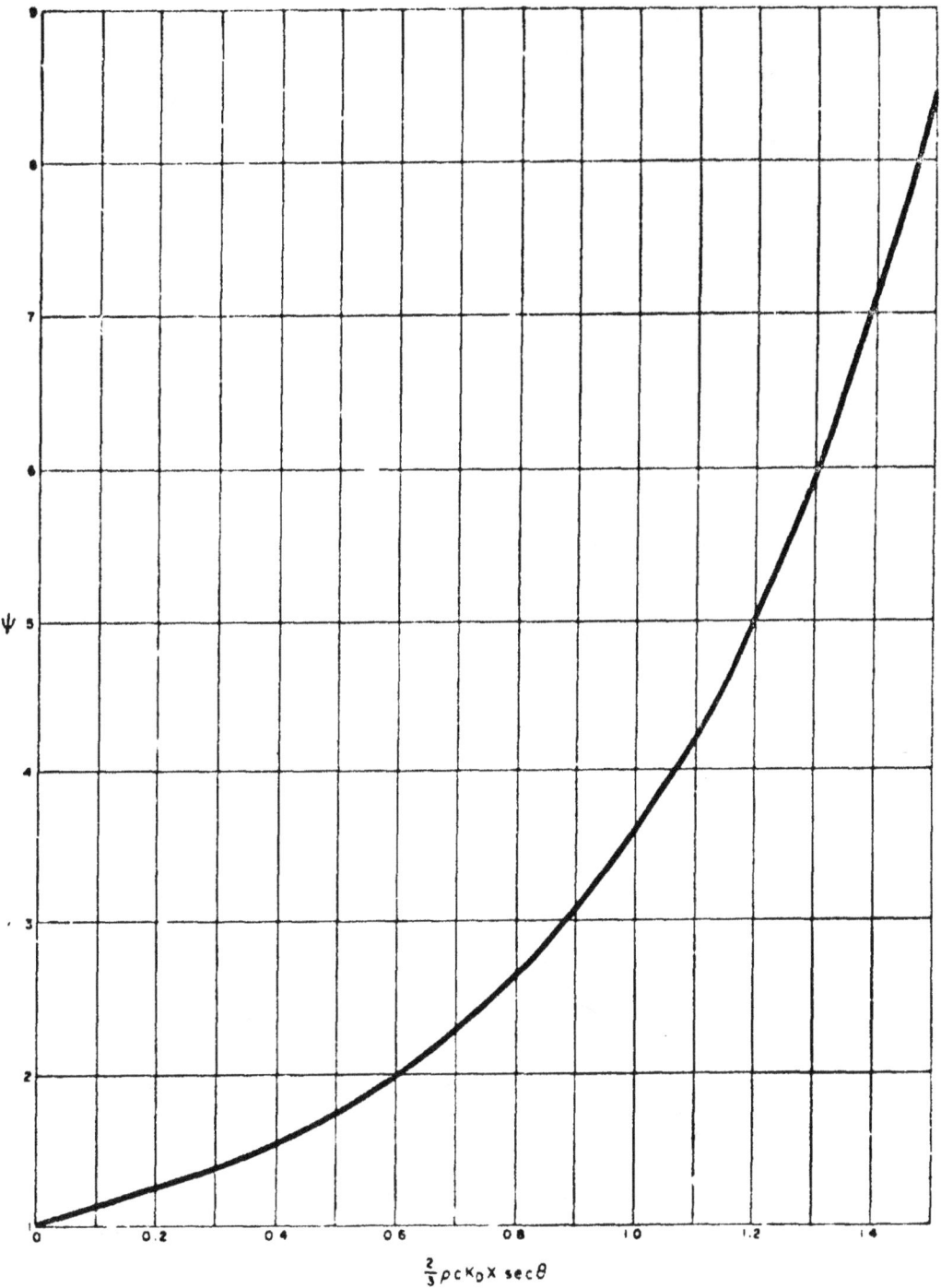

FIG. 4. Ballistic Drag Function.

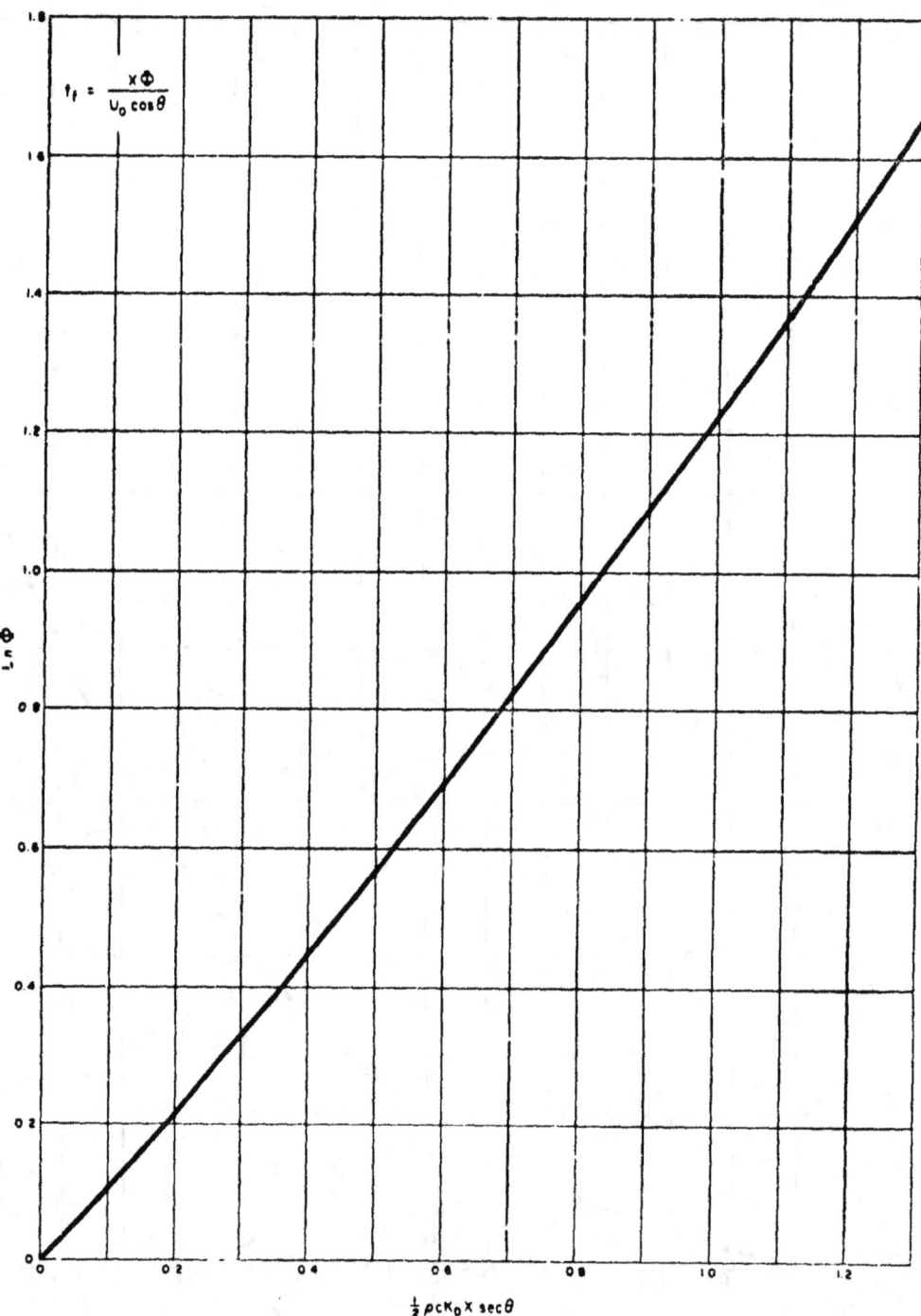

FIG. 5. Ballistic Drag Function.

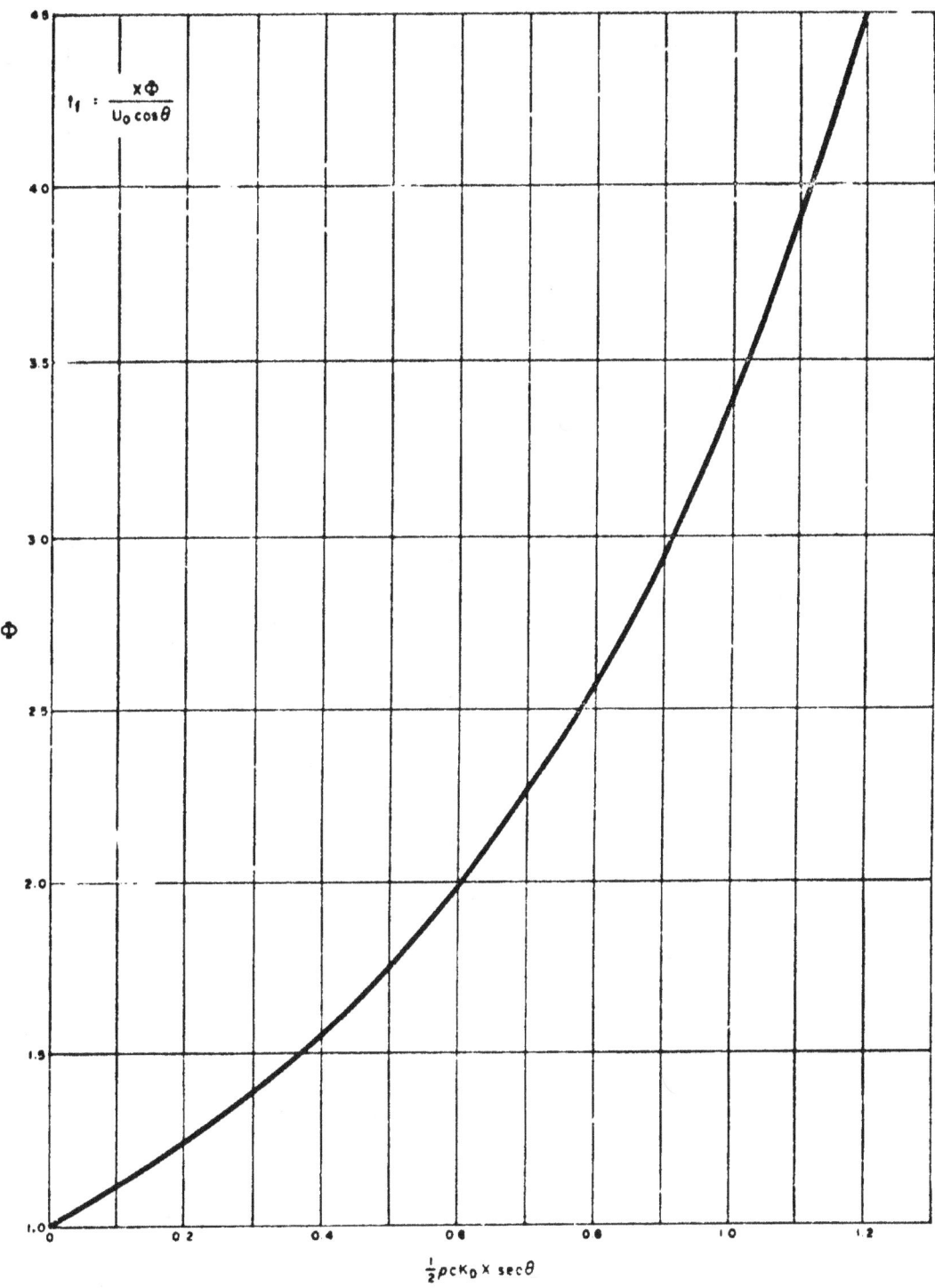

FIG. 6. Ballistic Drag Function.

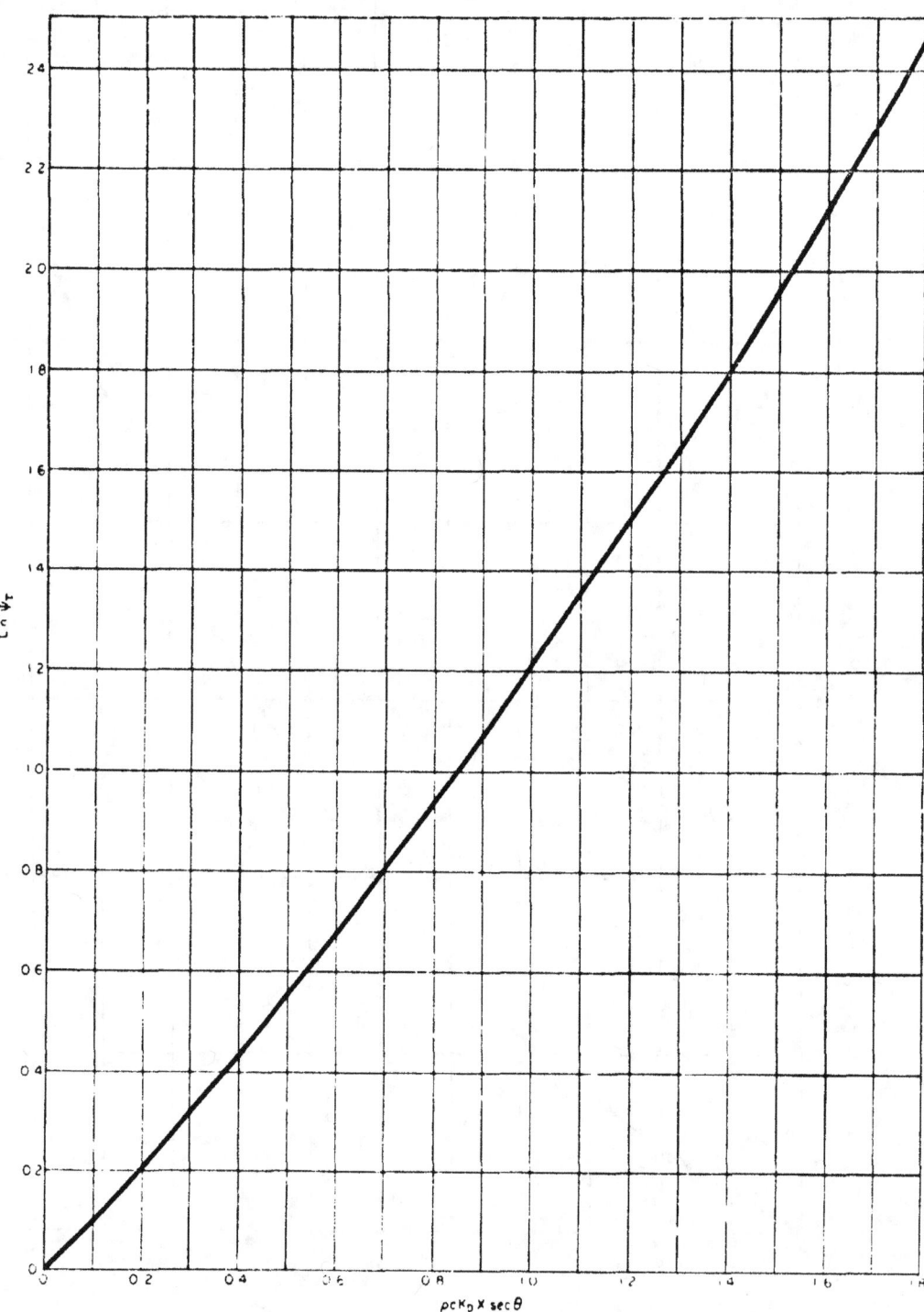

FIG. 7. Ballistic Drag Function.

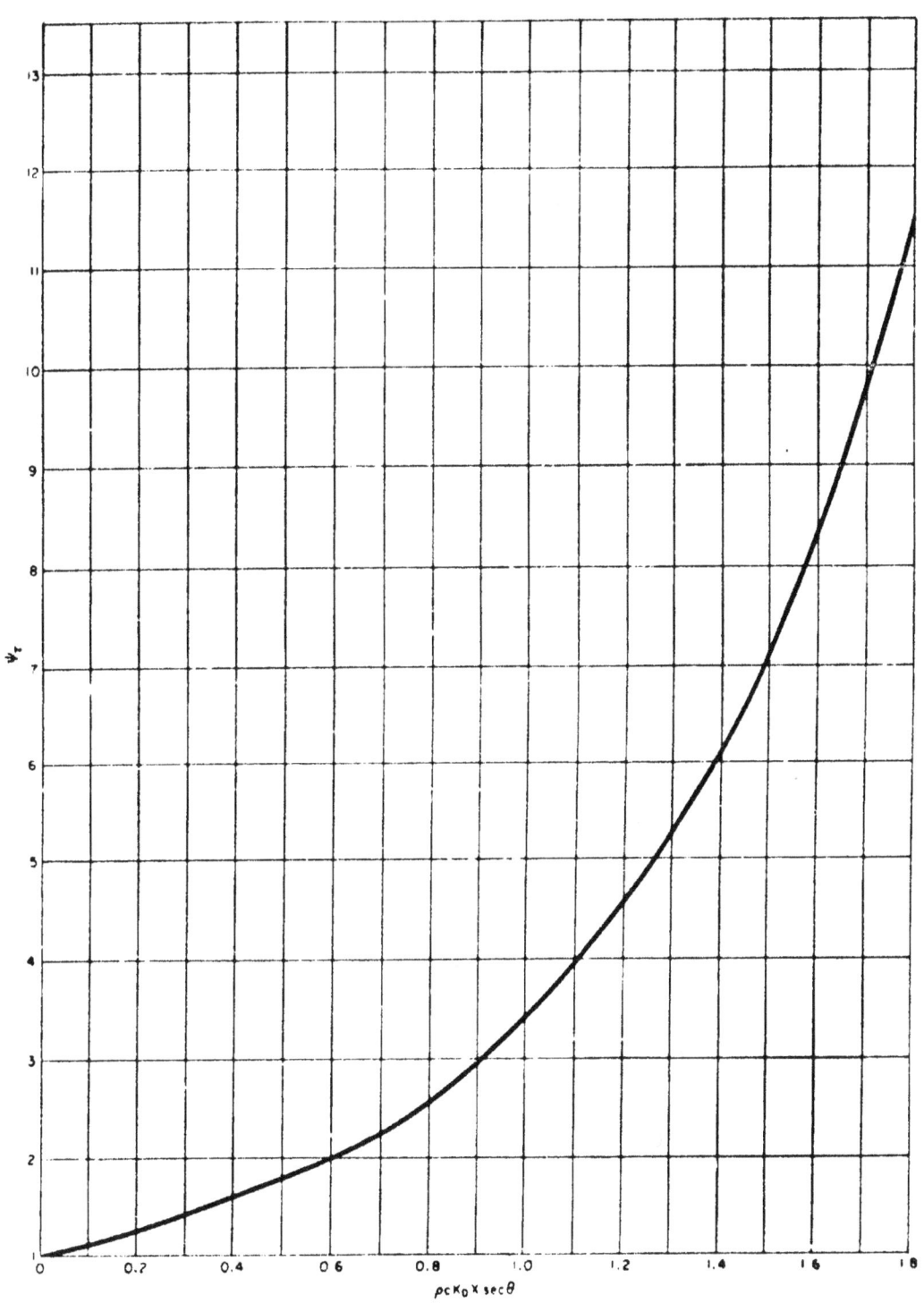

FIG. 8. Ballistic Drag Function.

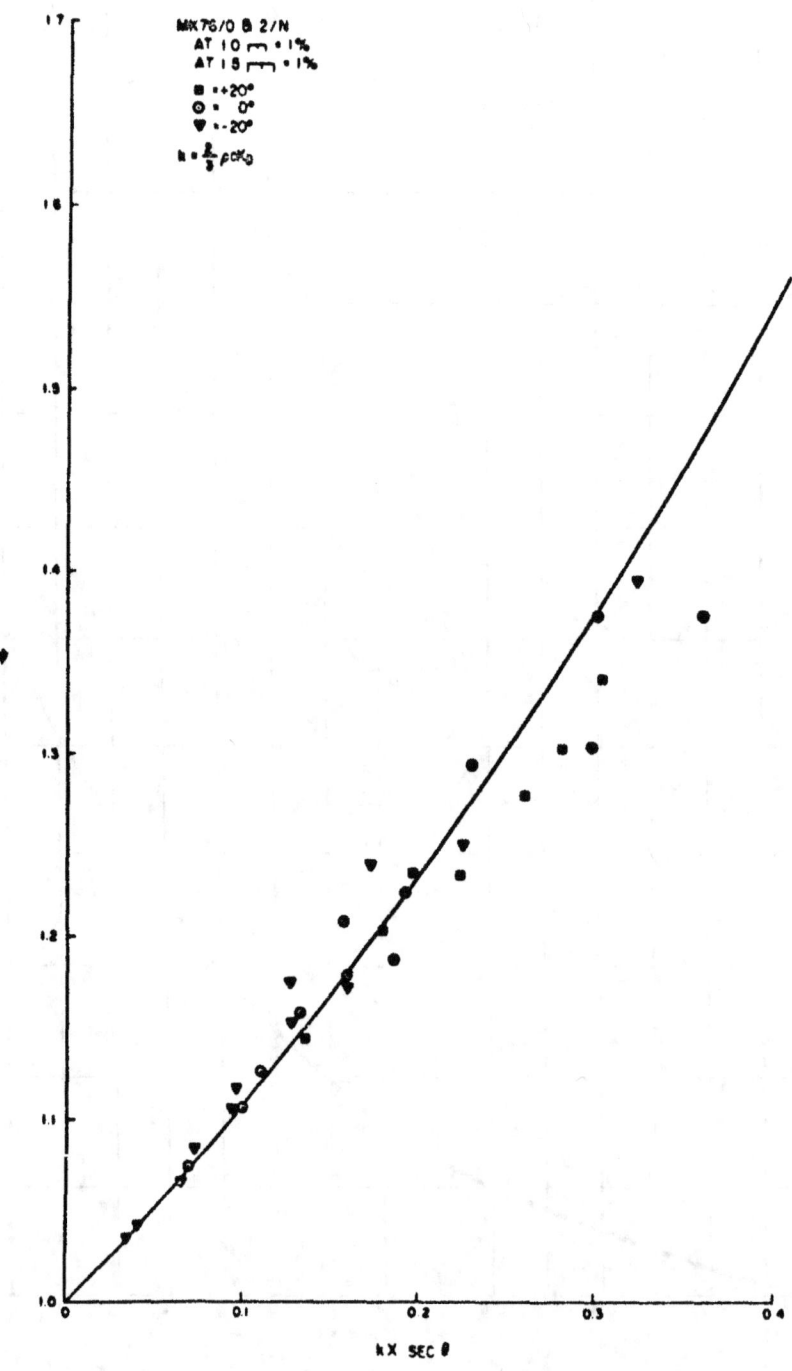

FIG. 9. Ballistic Drag Function.

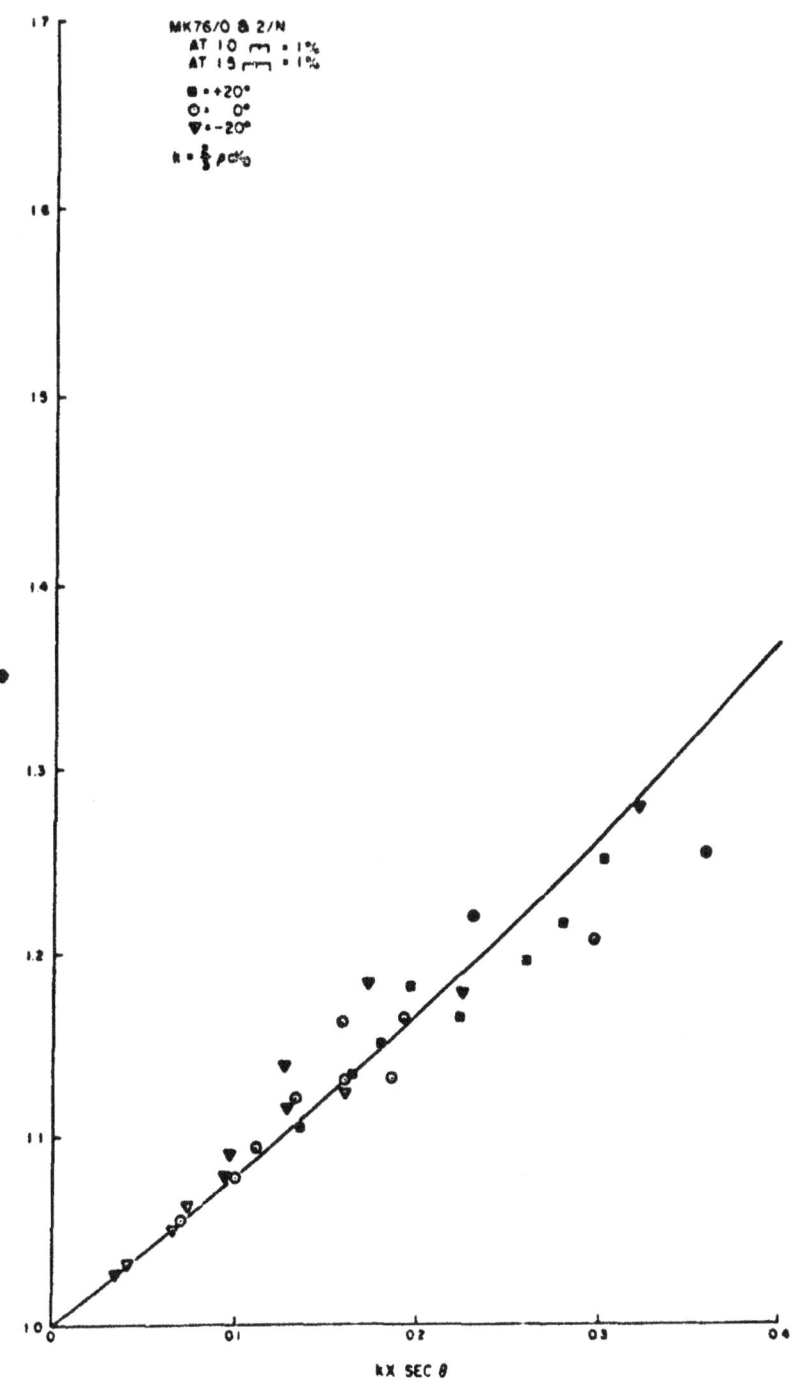

FIG. 10. Ballistic Drag Function.

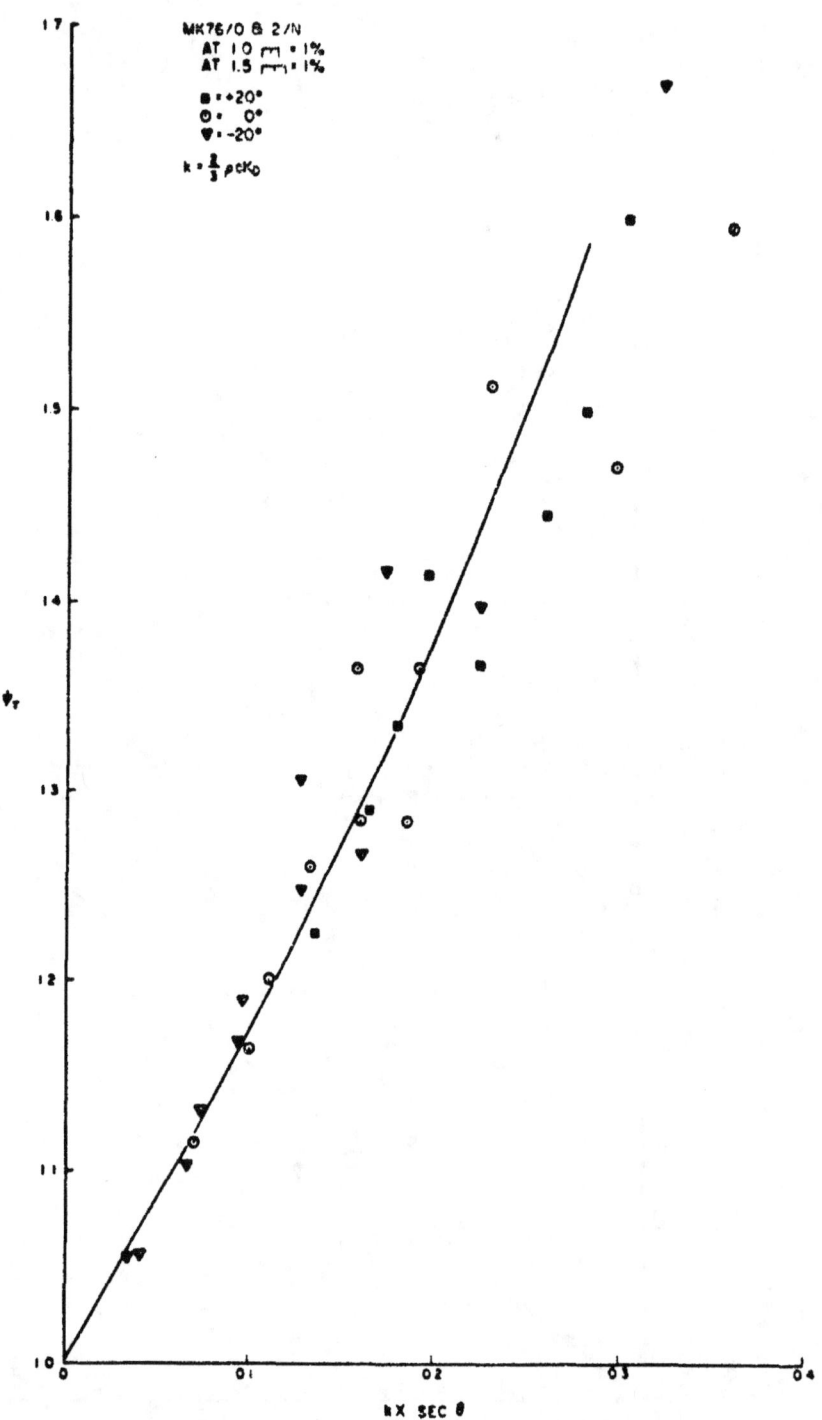

FIG. 11. Ballistic Drag Function.

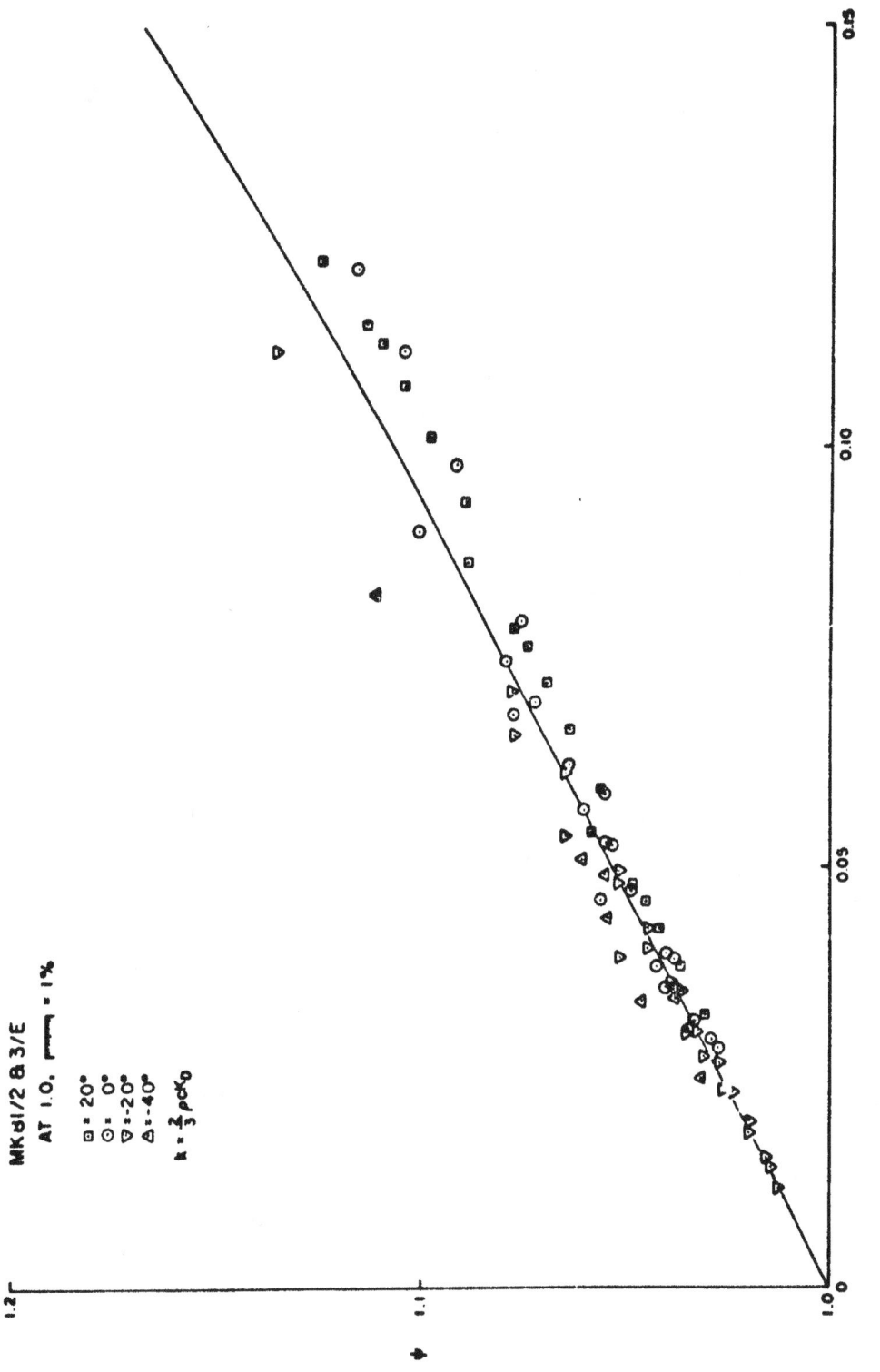

FIG. 12. Ballistic Drag Function.

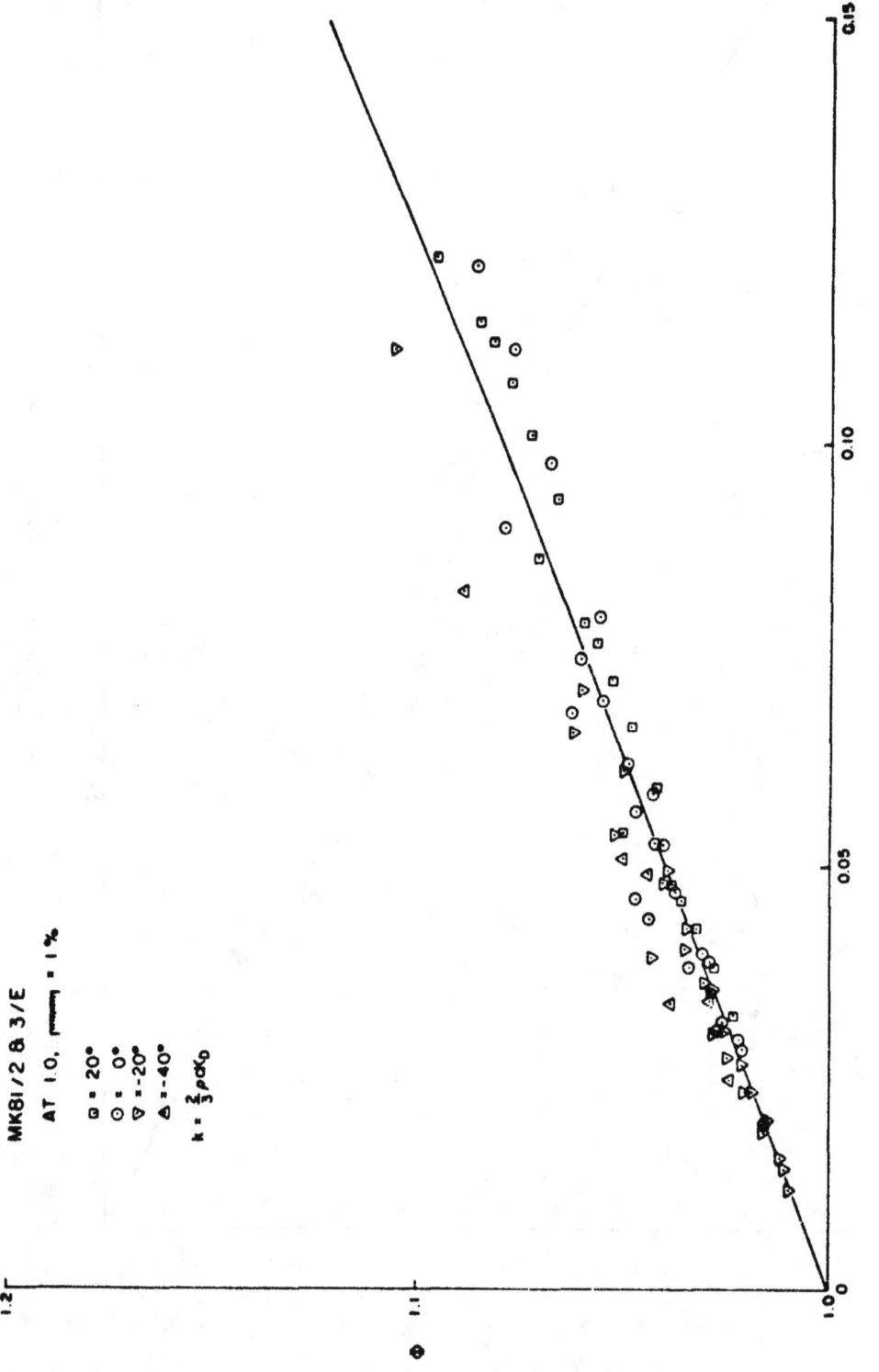

FIG. 13. Ballistic Drag Function.

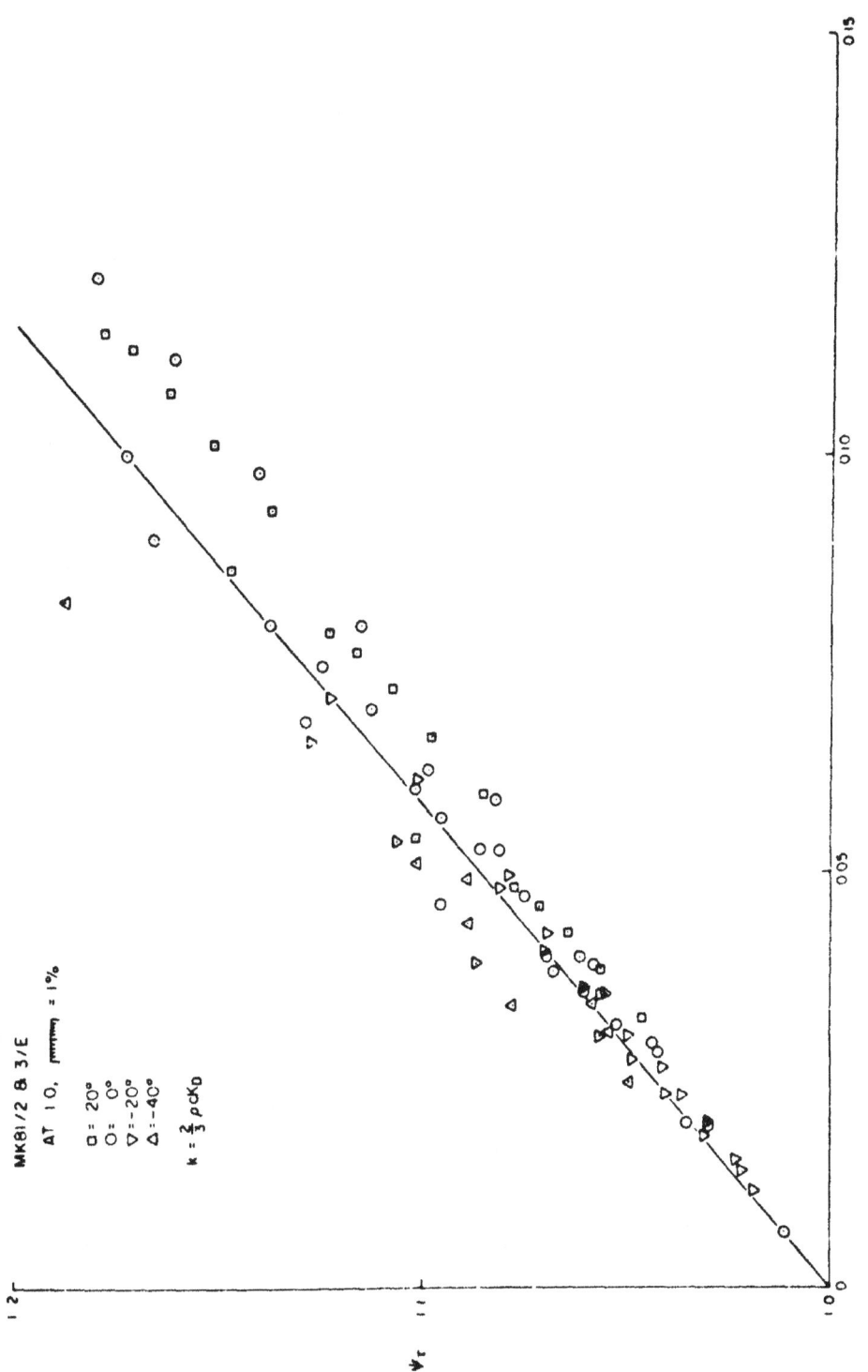

FIG. 14. Ballistic Drag Function.

NOTS TP 3902

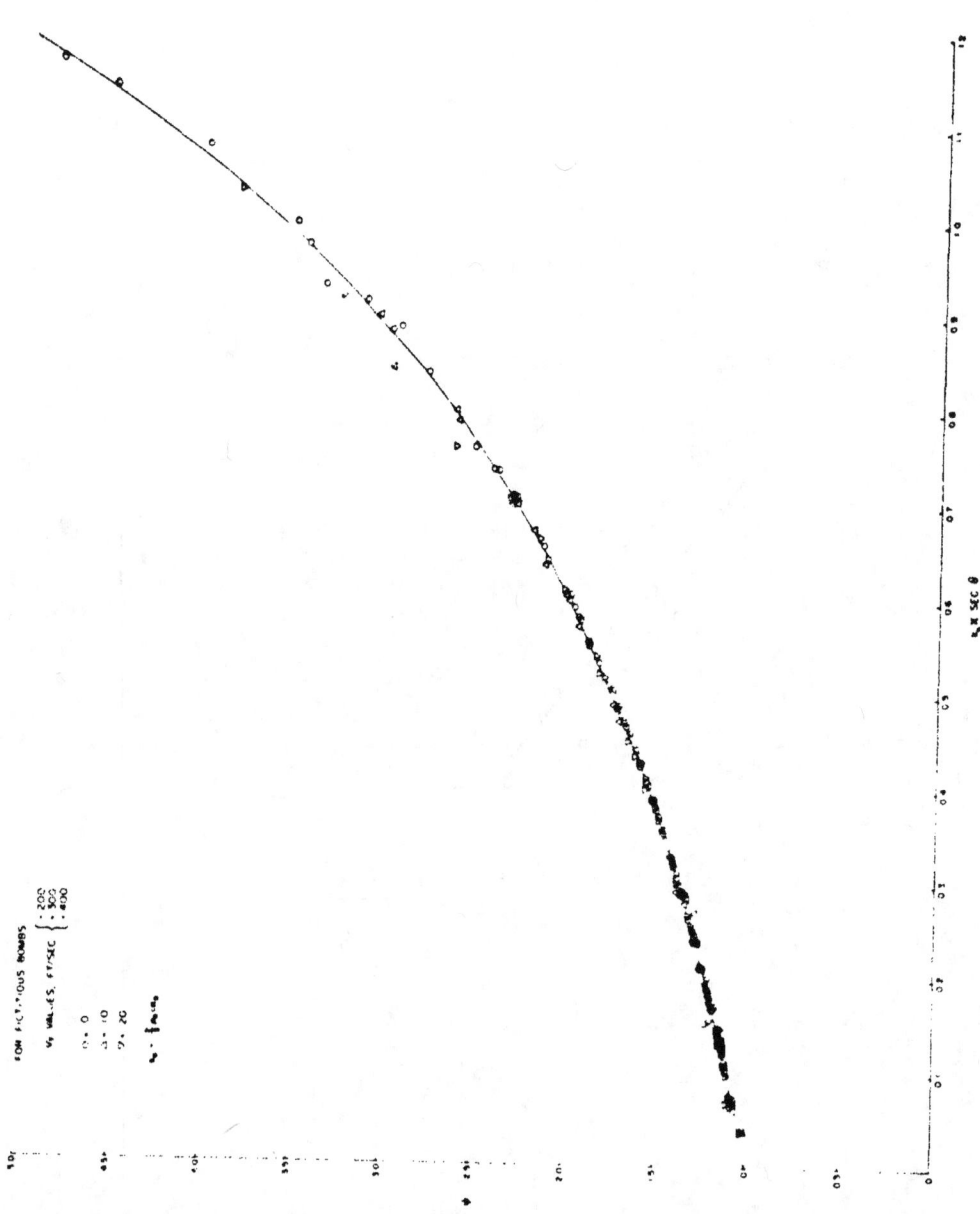

FIG. 15. Ballistic Drag Function.

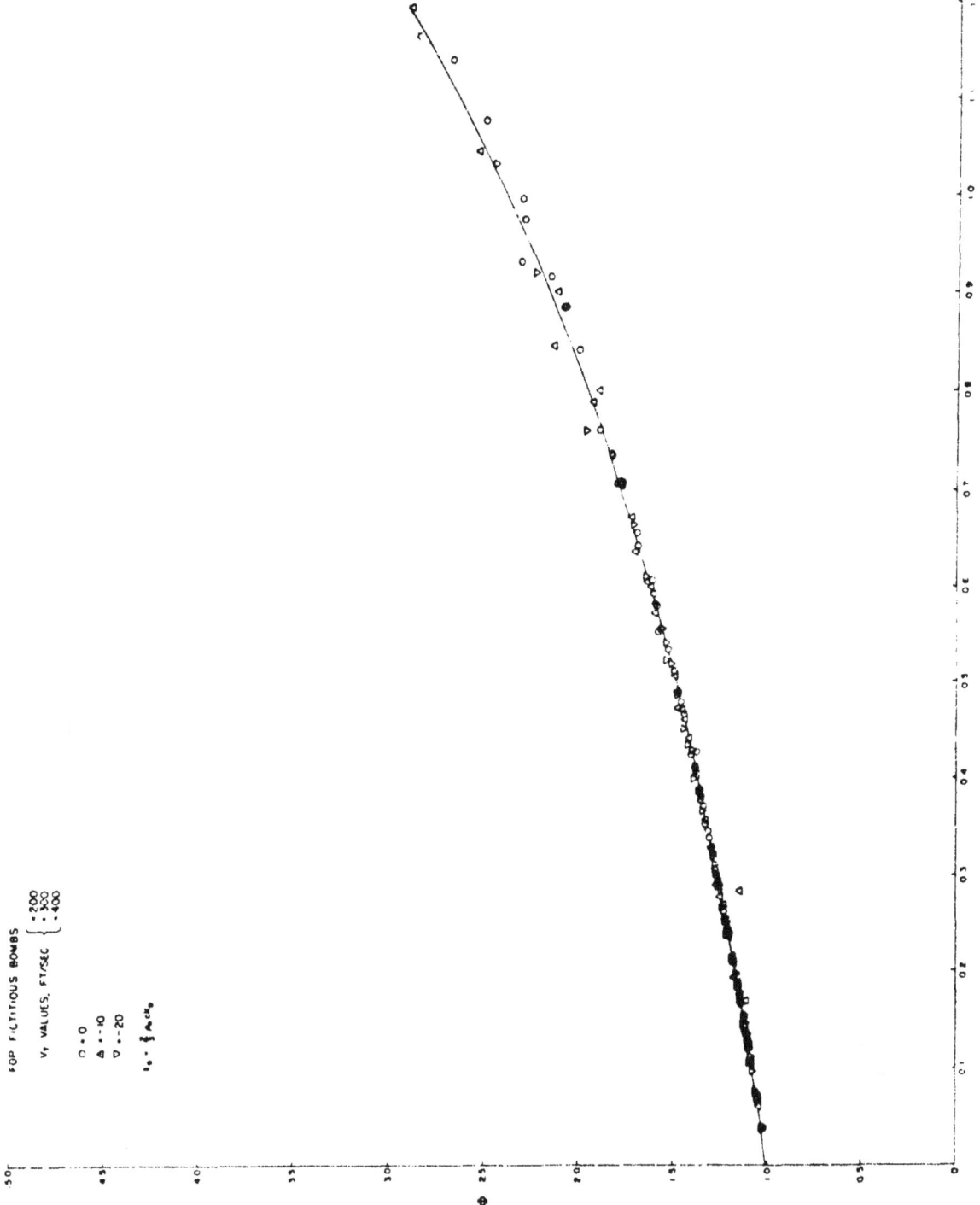

FIG. 16. Ballistic Drag Function.

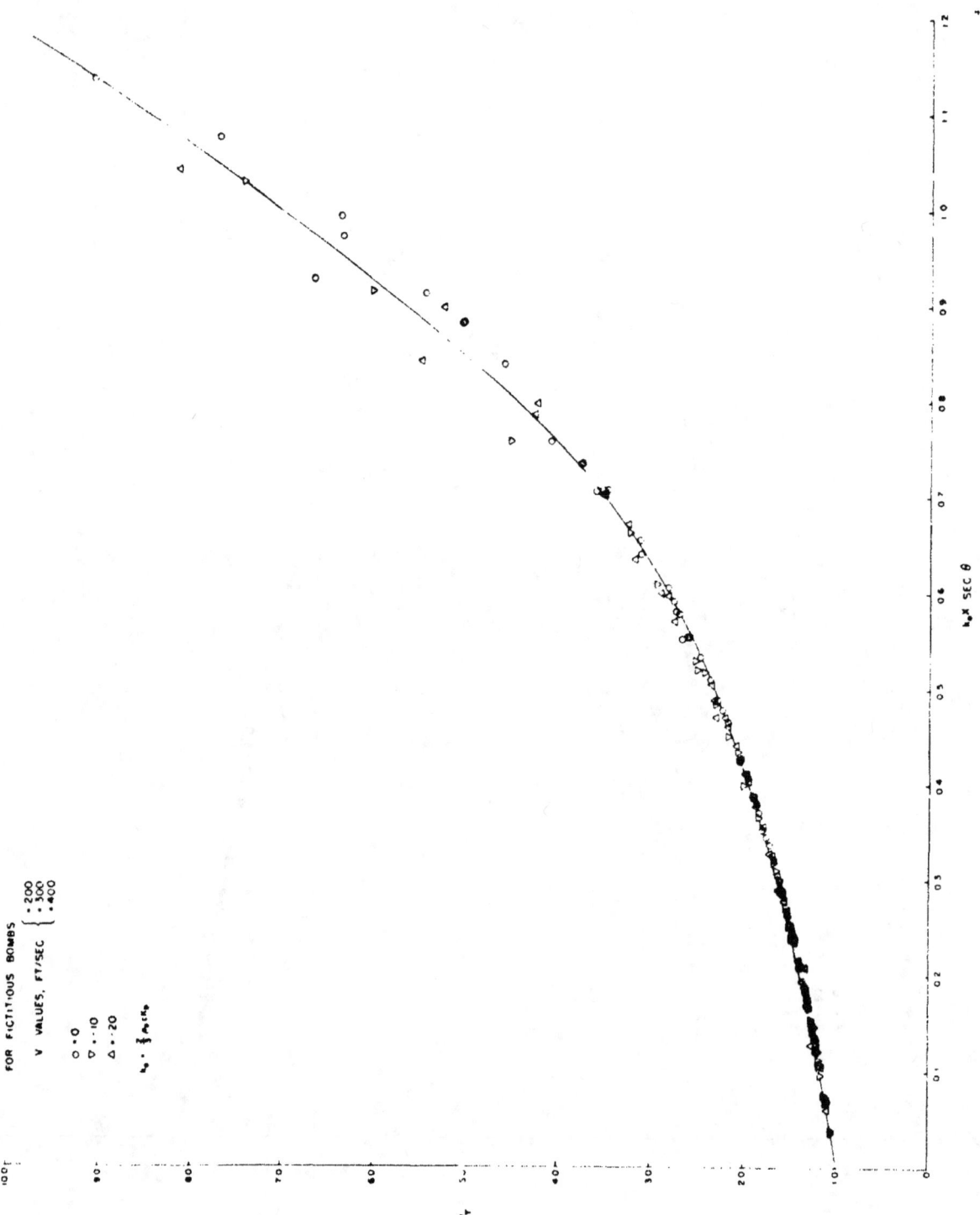

TABLE 2. Ballistic Function ψ and ln ψ Versus 2/3 ρc K_D x sec υ (kX sec υ)

$\frac{kX}{\cos υ}$	ψ	ln ψ	$\frac{kX}{\cos υ}$	ψ	ln ψ	$\frac{kX}{\cos υ}$	ψ	ln ψ	$\frac{kX}{\cos υ}$	ψ	ln ψ	$\frac{kX}{\cos υ}$	ψ	ln ψ
.00	1.000	.000	.30	1.380	.322	.60	2.003	.695	.90	3.070	1.121	1.20	5.012	1.611
.01	1.010	.010	.31	1.396	.334	.61	2.029	.708	.91	3.115	1.135	1.21	5.10	1.629
.02	1.020	.020	.32	1.413	.346	.62	2.055	.721	.92	3.161	1.150	1.22	5.20	1.647
.03	1.030	.030	.33	1.429	.357	.63	2.082	.734	.93	3.207	1.165	1.23	5.29	1.665
.04	1.041	.040	.34	1.446	.369	.64	2.110	.747	.94	3.255	1.180	1.24	5.39	1.683
.05	1.051	.050	.35	1.464	.381	.65	2.140	.761	.95	3.305	1.195	1.25	5.48	1.701
.06	1.062	.060	.36	1.481	.393	.66	2.170	.775	.96	3.355	1.211	1.26	5.58	1.719
.07	1.073	.070	.37	1.498	.404	.67	2.200	.789	.97	3.410	1.227	1.27	5.68	1.737
.08	1.084	.081	.38	1.516	.416	.68	2.233	.804	.98	3.467	1.243	1.28	5.78	1.754
.09	1.096	.092	.39	1.534	.428	.69	2.266	.818	.99	3.524	1.260	1.29	5.88	1.771
.10	1.107	.102	.40	1.552	.440	.70	2.299	.832	1.00	3.581	1.276	1.30	5.98	1.788
.11	1.119	.112	.41	1.570	.451	.71	2.332	.847	1.01	3.638	1.291	1.31	6.08	1.805
.12	1.131	.123	.42	1.589	.463	.72	2.366	.861	1.02	3.695	1.307	1.32	6.19	1.822
.13	1.143	.134	.43	1.608	.475	.73	2.400	.875	1.03	3.752	1.322	1.33	6.30	1.839
.14	1.156	.145	.44	1.627	.487	.74	2.434	.889	1.04	3.810	1.338	1.34	6.40	1.856
.15	1.168	.155	.45	1.647	.499	.75	2.470	.904	1.05	3.870	1.353	1.35	6.51	1.873
.16	1.181	.166	.46	1.667	.511	.76	2.507	.919	1.06	3.932	1.369	1.36	6.62	1.890
.17	1.194	.177	.47	1.688	.524	.77	2.542	.933	1.07	3.995	1.385	1.37	6.73	1.907
.18	1.207	.188	.48	1.709	.536	.78	2.578	.947	1.08	4.060	1.401	1.38	6.85	1.924
.19	1.220	.199	.49	1.731	.549	.79	2.614	.961	1.09	4.127	1.418	1.39	6.96	1.941
.20	1.234	.210	.50	1.754	.562	.80	2.652	.975	1.10	4.195	1.435	1.40	7.08	1.958
.21	1.248	.222	.51	1.777	.575	.81	2.690	.990	1.11	4.270	1.452	1.41	7.20	1.975
.22	1.262	.233	.52	1.801	.588	.82	2.728	1.004	1.12	4.345	1.469	1.42	7.33	1.993
.23	1.276	.244	.53	1.825	.602	.83	2.767	1.018	1.13	4.420	1.487	1.43	7.47	2.011
.24	1.290	.255	.54	1.849	.615	.84	2.805	1.032	1.14	4.500	1.504	1.44	7.61	2.030
.25	1.305	.266	.55	1.874	.628	.85	2.847	1.047	1.15	4.580	1.522	1.45	7.76	2.049
.26	1.320	.278	.56	1.900	.642	.86	2.892	1.062	1.16	4.662	1.539	1.46	7.91	2.068
.27	1.335	.289	.57	1.926	.655	.87	2.937	1.077	1.17	4.745	1.557	1.47	8.07	2.088
.28	1.350	.300	.58	1.952	.669	.88	2.980	1.092	1.18	4.832	1.575	1.49	8.38	2.126
.29	1.365	.311	.59	1.978	.682	.89	3.025	1.107	1.19	4.920	1.593	1.50	8.54	2.145

TABLE 3. Ballistic Function ϕ and $\ln \phi$ Versus $1/2 \rho c\, K_D \times \sec\theta$ ($3/4\, kX \sec\theta$)

$\frac{3/4\, kX}{\cos\theta}$	ϕ	$\ln \phi$	$\frac{3/4\, kX}{\cos\theta}$	ϕ	$\ln \phi$	$\frac{3/4\, kX}{\cos\theta}$	ϕ	$\ln \phi$	$\frac{3/4\, kX}{\cos\theta}$	ϕ	$\ln \phi$
.00	1.000	.000	.25	1.309	.269	.50	1.756	.563	.75	2.413	.881
.01	1.012	.012	.26	1.322	.279	.51	1.777	.575	.76	2.442	.893
.02	1.024	.024	.27	1.333	.291	.52	1.800	.588	.77	2.479	.908
.03	1.035	.034	.28	1.353	.302	.53	1.820	.599	.78	2.509	.920
.04	1.044	.043	.29	1.367	.313	.54	1.844	.612	.79	2.545	.934
.05	1.055	.054	.30	1.384	.325	.55	1.866	.624	.80	2.580	.948
.06	1.066	.064	.31	1.399	.336	.56	1.891	.637	.81	2.617	.962
.07	1.077	.074	.32	1.416	.348	.57	1.914	.649	.82	2.651	.975
.08	1.089	.085	.33	1.432	.359	.58	1.939	.662	.83	2.686	.988
.09	1.100	.095	.34	1.449	.371	.59	1.962	.674	.84	2.724	1.002
.10	1.112	.106	.35	1.467	.383	.60	1.990	.688	.85	2.762	1.016
.11	1.123	.116	.36	1.481	.393	.61	2.014	.700	.86	2.798	1.029
.12	1.135	.127	.37	1.498	.404	.62	2.040	.713	.87	2.838	1.043
.13	1.148	.138	.38	1.519	.418	.63	2.067	.726	.88	2.875	1.056
.14	1.159	.148	.39	1.536	.429	.64	2.092	.738	.89	2.915	1.070
.15	1.171	.158	.40	1.556	.442	.65	2.117	.750	.90	2.956	1.084
.16	1.185	.170	.41	1.571	.452	.66	2.145	.763	.91	2.995	1.097
.17	1.198	.181	.42	1.590	.464	.67	2.175	.777	.92	3.037	1.111
.18	1.212	.192	.43	1.613	.478	.68	2.201	.789	.93	3.077	1.124
.19	1.224	.202	.44	1.632	.490	.69	2.228	.801	.94	3.124	1.139
.20	1.237	.213	.45	1.652	.502	.70	2.259	.815	.95	3.168	1.153
.21	1.252	.225	.46	1.670	.513	.71	2.289	.828	.96	3.216	1.168
.22	1.266	.236	.47	1.692	.526	.72	2.321	.842	.97	3.258	1.181
.23	1.281	.248	.48	1.713	.538	.73	2.351	.855	.98	3.300	1.194
.24	1.294	.258	.49	1.733	.550	.74	2.382	.868	.99	3.347	1.208
									1.00	3.397	1.223
									1.01	3.445	1.237
									1.02	3.490	1.250
									1.03	3.543	1.265
									1.04	3.593	1.279
									1.05	3.644	1.293
									1.06	3.677	1.302
									1.07	3.751	1.322
									1.08	3.808	1.337
									1.09	3.857	1.350
									1.10	3.912	1.364
									1.11	3.971	1.379
									1.12	4.031	1.394
									1.13	4.088	1.408
									1.14	4.145	1.422
									1.15	4.212	1.438
									1.16	4.272	1.452
									1.17	4.332	1.466

TABLE 4. Ballistic Function ψ_T and $\ln \psi_T$ Versus $\rho c\, K_D \times \sec\theta$ $(3/2\, kX \sec\theta)$

$\frac{3/2\, kX}{\sec\theta}$	ψ_T	$\ln\psi_T$	$\frac{3/2\, kX}{\sec\theta}$	ψ_T	$\ln\psi_T$	$\frac{3/2\, kX}{\sec\theta}$	ψ_T	$\ln\psi_T$	$\frac{3/2\, kX}{\sec\theta}$	ψ_T	$\ln\psi_T$
.00	1.000	.000	.25	1.306	.267	.50	1.747	.558	1.00	3.380	1.218
.01	1.011	.011	.26	1.320	.278	.51	1.765	.568	1.01	3.432	1.233
.02	1.022	.022	.27	1.335	.289	.52	1.788	.581	1.02	3.483	1.248
.03	1.035	.034	.28	1.350	.300	.53	1.811	.594	1.03	3.532	1.262
.04	1.047	.046	.29	1.365	.311	.54	1.828	.603	1.04	3.585	1.277
.05	1.059	.057	.30	1.381	.323	.55	1.857	.619	1.05	3.636	1.291
.06	1.070	.068	.31	1.397	.334	.56	1.881	.632	1.06	3.684	1.304
.07	1.081	.078	.32	1.413	.346	.57	1.902	.643	1.07	3.743	1.320
.08	1.092	.088	.33	1.428	.356	.58	1.925	.655	1.08	3.792	1.333
.09	1.102	.097	.34	1.445	.368	.59	1.948	.667	1.09	3.850	1.348
.10	1.111	.105	.35	1.462	.380	.60	1.974	.680	1.10	3.904	1.362
.11	1.124	.117	.36	1.477	.390	.61	1.998	.692	1.11	3.963	1.377
.12	1.138	.129	.37	1.495	.402	.62	2.020	.703	1.12	4.023	1.392
.13	1.150	.140	.38	1.514	.415	.63	2.046	.716	1.13	4.084	1.407
.14	1.162	.150	.39	1.531	.426	.64	2.075	.730	1.14	4.145	1.422
.15	1.173	.160	.40	1.550	.438	.65	2.106	.745	1.15	4.208	1.437
.16	1.185	.170	.41	1.567	.449	.66	2.130	.756	1.16	4.263	1.450
.17	1.197	.180	.42	1.587	.462	.67	2.160	.770	1.17	4.328	1.465
.18	1.209	.190	.43	1.603	.472	.68	2.186	.782	1.18	4.393	1.480
.19	1.221	.200	.44	1.624	.485	.69	2.219	.797	1.19	4.459	1.495
.20	1.235	.211	.45	1.645	.498	.70	2.243	.808	1.20	4.527	1.510
.21	1.249	.222	.46	1.664	.509	.71	2.273	.821	1.21	4.595	1.525
.22	1.262	.233	.47	1.684	.521	.72	2.300	.833	1.22	4.665	1.540
.23	1.278	.245	.48	1.704	.533	.73	2.335	.848	1.23	4.735	1.555
.24	1.290	.255	.49	1.725	.545	.74	2.363	.860	1.24	4.807	1.570

Additional middle section:

$\frac{3/2\, kX}{\sec\theta}$	ψ_T	$\ln\psi_T$
.75	2.396	.874
.76	2.428	.887
.77	2.460	.900
.78	2.494	.914
.79	2.529	.928
.80	2.565	.942
.81	2.599	.955
.82	2.635	.969
.83	2.670	.982
.84	2.710	.997
.85	2.743	1.009
.86	2.782	1.023
.87	2.821	1.037
.88	2.861	1.051
.89	2.895	1.063
.90	2.939	1.078
.91	2.980	1.092
.92	3.050	1.115
.93	3.065	1.120
.94	3.105	1.133
.95	3.152	1.148
.96	3.196	1.162
.97	3.238	1.175
.98	3.287	1.190
.99	3.333	1.204

TABLE 4. Ballistic Function ψ_T and $\ln \psi_T$ Versus $\rho c\, K_D \times \sec \upsilon$ ($3/2\, kX \sec \upsilon$) (Cont'd.)

$3/2\, kX \sec \upsilon$	ψ_T	$\ln \psi_T$	$3/2\, kX \sec \upsilon$	ψ_T	$\ln \psi_T$	$3/2\, kX \sec \upsilon$	ψ_T	$\ln \psi_T$	$3/2\, kX \sec \upsilon$	ψ_T	$\ln \psi_T$
1.25	4.884	1.586	1.50	7.171	1.970	1.75	10.687	2.369	2.00	15.991	2.772
1.26	4.953	1.600	1.51	7.301	1.988	1.76	10.859	2.385			
1.27	5.038	1.617	1.52	7.404	2.002	1.77	11.045	2.402			
1.28	5.114	1.632	1.53	7.531	2.019	1.78	11.223	2.418			
1.29	5.191	1.647	1.54	7.645	2.034	1.79	11.404	2.434			
1.30	5.265	1.661	1.55	7.768	2.050	1.80	11.600	2.451			
1.31	5.349	1.677	1.56	7.901	2.067	1.81	11.787	2.467			
1.32	5.430	1.692	1.57	8.020	2.082	1.82	11.989	2.484			
1.33	5.501	1.705	1.58	8.150	2.098	1.83	12.170	2.499			
1.34	5.596	1.722	1.59	8.265	2.112	1.84	12.367	2.515			
1.35	5.686	1.738	1.60	8.398	2.128	1.85	12.579	2.532			
1.36	5.766	1.752	1.61	8.542	2.145	1.86	12.769	2.547			
1.37	5.859	1.768	1.62	8.671	2.160	1.87	12.975	2.563			
1.38	5.948	1.783	1.63	8.811	2.176	1.88	13.171	2.578			
1.39	6.050	1.800	1.64	8.953	2.192	1.89	13.397	2.595			
1.40	6.135	1.814	1.65	9.088	2.207	1.90	13.626	2.612			
1.41	6.234	1.830	1.66	9.235	2.223	1.91	13.846	2.628			
1.42	6.341	1.847	1.67	9.393	2.240	1.92	14.083	2.645			
1.43	6.430	1.861	1.68	9.545	2.256	1.93	14.325	2.662			
1.44	6.540	1.878	1.69	9.699	2.272	1.94	14.556	2.678			
1.45	6.639	1.893	1.70	9.865	2.289	1.95	14.791	2.694			
1.46	6.740	1.908	1.71	10.004	2.303	1.96	15.029	2.710			
1.47	6.841	1.923	1.72	10.186	2.321	1.97	15.256	2.725			
1.48	6.952	1.939	1.73	10.350	2.337	1.98	15.518	2.742			
1.49	7.057	1.954	1.74	10.517	2.353	1.99	15.753	2.757			

One of the nomographs uses the strictly empirical equation:

$$t_f = \sqrt{\frac{2Z}{g}} \, e^{(k_{Zt})Z^{(1/3.5)}} \qquad (60)$$

Figure 18 gives a graph of k_{Zt} versus terminal velocity.

4. Partial Derivative Equations

The partial derivative equations of this section were obtained from the closed trajectory equations of Section 3. For an initial set of release conditions and the resulting trajectory parameters, the equations will give the change in the dependent variable due to an incremental change in the independent variable from its value in the initial set, all other release variables remaining constant. Where possible, several equivalent forms of the equations are given.

For simplicity the assumption is made that $\ln \psi = (2/3) \rho c K_D X \sec \theta$, and $\psi = \exp (2/3) \rho c K_D X \sec \theta$. For very high drag weapons this approximation for the more exact expressions given by eq. 52, 54, and 56, will result in some error, but this is usually negligible for the purpose for which these equations are normally used.

(1) Dependent variable X, horizontal range

$$\frac{\partial X}{\partial Z} = \frac{2U_o^2 \cos^2 \theta}{gX\psi \, (2 + \ln\psi) - 2U_o^2 \sin \theta \cos \theta} \qquad (61a)$$

$$= \frac{X}{Z + (Z + X \tan \theta)(1 + \ln\psi)} \qquad (61b)$$

$$= \frac{1}{\tan \phi + (\tan \phi + \tan \theta)(1 + \ln\psi)} \qquad (61c)$$

$$= - \cot \tau \qquad (61d)$$

$$\frac{\partial X}{\partial U_o} = \frac{2}{U_o} \, \frac{gX^2 \psi}{gX\psi (2 + \ln\psi) - 2U_o^2 \sin \theta \cos \theta} \qquad (62a)$$

$$= \frac{2X}{U_o} \, \frac{Z/X + \tan \theta}{(Z/X + \tan \theta)(2 + \ln\psi) - \tan \theta} \qquad (62b)$$

35

$$= \frac{2X}{U_o} \frac{\tan \phi + \tan \theta}{(\tan \phi + \tan \theta)(2 + \ln\psi) - \tan \theta} \tag{62c}$$

$$\frac{\partial X}{\partial \theta} = \frac{2U_o^2 - gX\psi \tan \theta (2 + \ln\psi)}{gX (2 + \ln \psi) - 2U_o^2 \sin \theta \cos \theta} \tag{63a}$$

$$= X \frac{\sec^2 \theta - \tan \theta (Z/X + \tan \theta)(2 + \ln\psi)}{(Z/X + \tan \theta) 2 + \ln\psi) - \tan} \tag{63b}$$

$$= X \frac{\sec^2 \theta - \tan \theta (\tan \phi + \tan \theta)(2 + \ln\psi)}{(\tan \phi + \tan \theta)(2 + \ln\psi) - \tan \theta} \tag{63c}$$

$$= - X (\cot \tau + \tan \theta) \tag{63d}$$

$$\frac{\partial X}{\partial (cK_D)} = - \frac{1}{cK_D} \frac{gX^2 \psi \ln\psi}{gX\psi (2 + \ln\psi) - 2U_o^2 \sin \theta \cos \theta} \tag{64a}$$

$$= - \frac{X}{cK_D} \frac{(\tan \phi + \tan \theta) \ln\psi}{\tan \phi + (\tan \theta + \tan \theta)(1 + \ln\psi)} \tag{64b}$$

$$= \frac{X}{cK_D} \cot \tau_1 (\tan \phi + \tan \theta) \ln\psi \tag{64c}$$

(2) Dependent variable t, time of fall

$$\frac{\partial t}{\partial Z} = - \frac{\psi^{3/4}}{U_o \cos \theta} \{1 + (3/4) \ln\psi\} \cot \tau \tag{65a}$$

$$= - \frac{t}{X} \{1 + (3/4) \ln\psi\} \cot \tau \tag{65b}$$

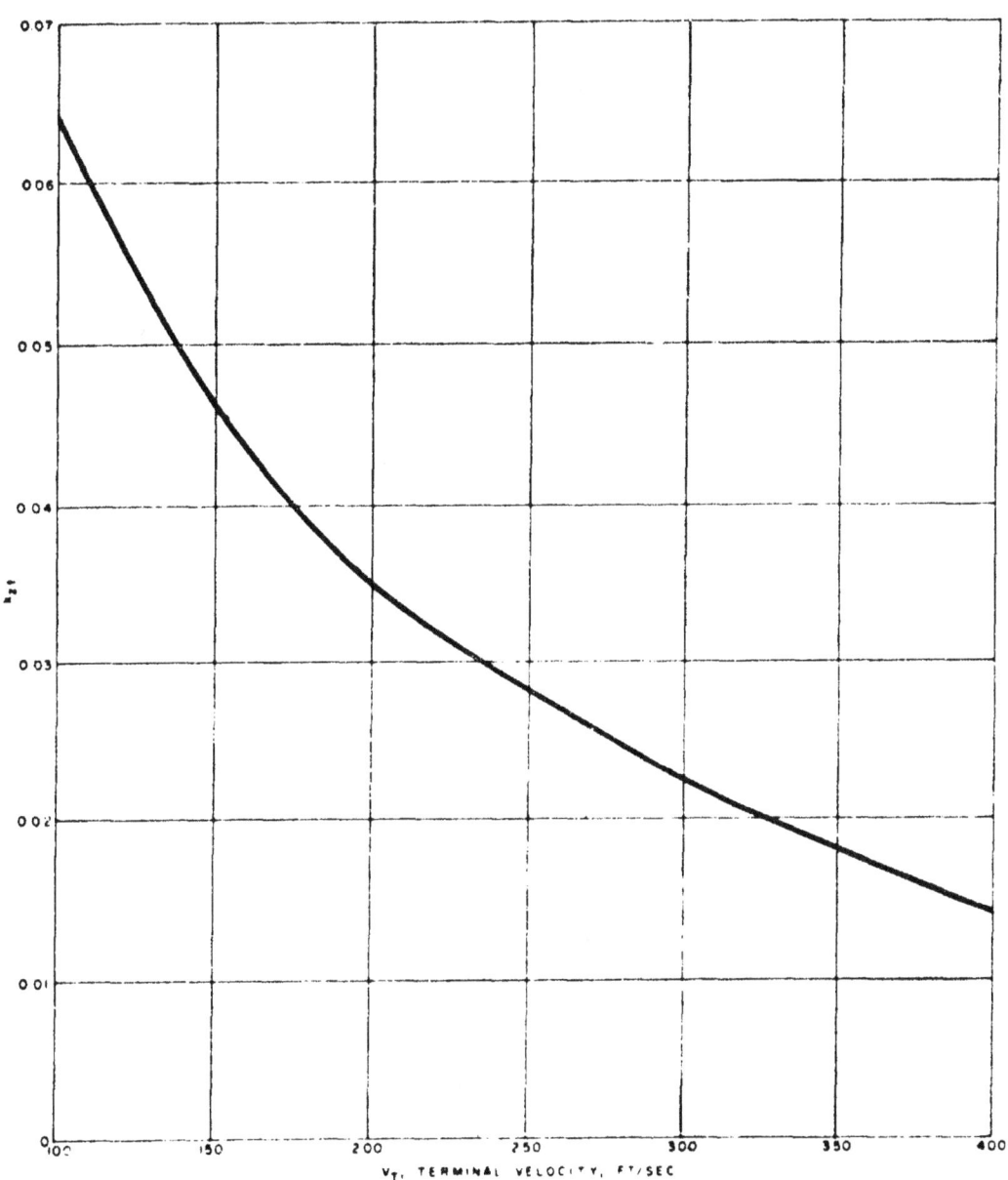

Fig. 18. Ballistic Drag Function.

$$\frac{\partial t}{\partial U_o} = - \frac{X\psi^{3/4}}{U_o^2 \cos\theta} \left[1 + 2\{1 + (3/4)\ln\psi\}(\tan\phi + \tan\theta)\cot\tau\right] \quad (66a)$$

$$= -\frac{t}{U_o}\left[1 + 2\{1 + (3/4)\ln\psi\}(\tan\phi + \tan\theta)\cot\tau\right] \quad (66b)$$

$$\frac{\partial t}{\partial \theta} = -t\cot\tau\{1 + (3/4)\ln\psi\} \quad (67)$$

$$\frac{\partial t}{\partial(cK_D)} = 3/4 \frac{t}{cK_D} \ln\psi \quad (68)$$

(3) Dependent variable τ, impact angle.

$$\frac{\partial \tau}{\partial U_o} = \frac{2\cos^2\tau}{U_o}(\tan\theta - \tan\tau)\left[1 - \cot\tau(\tan\phi + \tan\theta)\{1 + (3/2)\ln\psi\}\right] \quad (69)$$

$$\frac{\partial \tau}{\partial \theta} = \cot\tau\tan\theta\{1 + (3/4)\ln\psi\} \quad (70a)$$

$$= -\frac{\tan\theta\{1 + (3/2)\ln\psi\}}{\tan\phi + (\tan\phi + \tan\theta)(1 + \ln\psi)} \quad (70b)$$

$$\frac{\partial \tau}{\partial Z} = -\frac{\cos^2\tau \cot\tau^2 g\psi^{3/4}\{1 + (3/4)\ln\psi\}}{U_o^2 \cos^2\theta} \quad (71a)$$

$$= \cos^2\tau \cot\tau (\tan\theta - \tan\tau)\{1 + (3/4)\ln\psi\}/X \quad (71b)$$

$$\frac{\partial \tau}{\partial(cK_D)} = -\frac{\cos^2\tau}{cK_D}\left(\frac{gX\psi}{2U_o^2 \cos^2\theta}\right)\ln\psi \ (3 + \ln\psi) \quad (72a)$$

$$= \frac{\cos^2 \tau}{cK_D} (\tan \tau + \tan \theta + 2 \tan \phi)(3 + \ln \psi) \tag{72b}$$

$$= \frac{3 \cot \tau_1}{2 \, cK_D} \ln \psi \tag{72c}$$

(4) Dependent variable V, true airspeed

$$\frac{\partial U}{\partial \theta} = -U_o \sec \tau \sin \theta \psi^{-3/2} \left[\{1 + (3/2) \ln \psi\} \right] \tag{73a}$$

$$= -U \tan \theta \sec \tau \left[\{1 + (3/2) \ln \psi\} \right] \tag{73b}$$

$$\frac{\partial U}{\partial U_o} = \frac{\cos \theta}{\cos \tau_1} \psi^{-3/2} \tag{74a}$$

$$= \frac{U}{U_o} \tag{74b}$$

$$\frac{\partial U}{\partial Z} = -3/2 \left(\frac{U}{X} \cot \tau_1 \ln \psi \right) \tag{75}$$

$$\frac{\partial U}{\partial (cK_D)} = -\frac{3 \, U_o \cos \theta \psi^{-3/2}}{2 \, cK_D \cos \tau} \ln \psi \tag{76a}$$

$$= \frac{3 \, U \ln \psi}{2 \, cK_D} \tag{76b}$$

III. NUMERICAL DATA

A. CONVERSION FACTORS AND CONSTANTS

Various constants: π = 3.14159265
                        e = 2.71828183

Measures of length:

     1 meter = 1.0936 yards = 3.2808 feet = 39.3700 inches
     1 foot = 0.30480 meter
     1 inch = 2.5400 centimeter
     1 mile = 1.60935 kilometer
     1 kilometer = 0.62137 mile
     1 nautical mile = 6,076.1 feet
     1 mile = 0.86898 nautical mile

Measures of velocity:

     1 foot per second = 0.5925 knots
     1 knot = 1.6878 feet per second

Measures of pressure:

     1 pound per square foot = 0.01414 inches Hg at 32°F
                                 = 4.725×10^{-4} atmosphere
                                 = 0.006944 pounds per square inch
     1 millibar = 2.089 pounds per square foot
     1 atmosphere = 29.92 inches of Hg at 32°F

Angular measure:

     1 degree = 0.01745 radians = 0.002778 revolutions
     1 radian = 57.296 degrees

B. AIR DATA TABLE

Several parameters of interest are tabulated in Table 5 with respect to altitude. The values given here are in agreement with the values defined by ICAO. The following quantities are listed herein:

h = altitude in feet
ρ = density in lb/ft^3
ρ/ρ_0 = ratio of density at given altitude to density at zero altitude
P_s = the "standard" pressure for that altitude given in lb/ft^2
P_s/P_{s0} = the ratio of pressure at given altitude to pressure at zero altitude
C_s = speed of sound at given altitude
Conversion factor = Mach no./1000 fps
U_t/U_i = ratio of true velocity to indicated velocity
T = standard temperature in °F

TABLE 5. Air Data

h ft	$\rho \times 10^{-2}$ lb/ft^3	$\frac{\rho}{\rho_o}$	P_s lb/ft^2	$\frac{P_s}{P_{so}}$	C_s fps	$\frac{\text{Mach no.}}{1000\text{ fps}}$	U_t/U_i	T °F
0	7.648	1.0000	2116	1.000	1117	.8953	1.0000	59.00
200	7.603	0.9942	2101	0.9928	1116	.8961	1.0029	58.29
400	7.558	0.9884	2086	0.9856	1115	.8969	1.0059	57.57
600	7.514	0.9826	2071	0.9785	1115	.8969	1.0089	56.86
800	7.470	0.9768	2056	0.9714	1114	.8977	1.0118	56.15
1000	7.426	0.9711	2041	0.9644	1113	.8985	1.0148	55.43
1200	7.383	0.9654	2026	0.9574	1112	.8993	1.0178	54.72
1400	7.339	0.9597	2011	0.9504	1112	.8993	1.0208	54.01
1600	7.296	0.9540	1997	0.9435	1111	.9001	1.0238	53.29
1800	7.253	0.9484	1982	0.9366	1110	.9009	1.0268	52.58
2000	7.210	0.9428	1968	0.9298	1109	.9017	1.0299	51.87
2200	7.167	0.9372	1953	0.9230	1108	.9025	1.0330	51.15
2400	7.125	0.9316	1939	0.9163	1108	.9025	1.0361	50.44
2600	7.082	0.9261	1925	0.9095	1107	.9033	1.0391	49.73
2800	7.040	0.9206	1911	0.9029	1106	.9042	1.0422	49.02
3000	6.998	0.9151	1897	0.8962	1105	.9050	1.0454	48.30
3200	6.957	0.9097	1883	0.8896	1104	.9058	1.0485	47.59
3400	6.915	0.9042	1869	0.8831	1104	.9058	1.0516	46.88
3600	6.874	0.8938	1855	0.8766	1103	.9066	1.0548	46.16
3800	6.833	0.8934	1841	0.8701	1102	.9074	1.0580	45.45
4000	6.792	0.8881	1828	0.8637	1101	.9083	1.0611	44.74
4200	6.751	0.8828	1814	0.8573	1101	.9083	1.0643	44.02
4400	6.710	0.8774	1801	0.8509	1100	.9091	1.0676	43.31
4600	6.670	0.8722	1787	0.8446	1099	.9099	1.0708	42.60
4800	6.630	0.8669	1774	0.8383	1098	.9107	1.0740	41.88
5000	6.590	0.8617	1761	0.8320	1098	.9107	1.0773	41.17
5200	6.550	0.8565	1748	0.8258	1097	.9115	1.0806	40.46
5400	6.510	0.8513	1735	0.8197	1096	.9123	1.0838	39.74
5600	6.471	0.8461	1722	0.8135	1095	.9132	1.0872	39.03
5800	6.431	0.8410	1709	0.8074	1094	.9141	1.0905	38.32
6000	6.392	0.8359	1696	0.8014	1094	.9141	1.0938	37.60
6200	6.353	0.8308	1683	0.7954	1093	.9149	1.0971	36.89
6400	6.315	0.8257	1670	0.7894	1092	.9158	1.1005	36.18
6600	6.276	0.8207	1658	0.7834	1091	.9166	1.1039	35.46
6800	6.238	0.8156	1645	0.7775	1090	.9174	1.1073	34.75

TABLE 5. (Cont'd)

h ft	$\rho \times 10^{-2}$ lb/ft³	$\frac{\rho}{\rho_o}$	p_s lb/ft²	$\frac{P_s}{P_{so}}$	C_s fps	$\frac{\text{Mach no.}}{1000 \text{ fps}}$	U_t/U_i	T °F
7000	6.199	0.8106	1633	0.7716	1090	.9174	1.1107	34.04
7200	6.161	0.8057	1621	0.7658	1089	.9183	1.1141	33.32
7400	6.124	0.8007	1608	0.7600	1088	.9191	1.1177	32.61
7600	6.086	0.7958	1596	0.7542	1087	.9200	1.1210	31.90
7800	6.048	0.7909	1584	0.7485	1086	.9208	1.1245	31.18
8000	6.011	0.7860	1572	0.7428	1086	.9208	1.1279	30.47
8200	5.974	0.7812	1560	0.7371	1085	.9217	1.1314	29.76
8400	5.937	0.7763	1548	0.7315	1084	.9225	1.1349	29.04
8600	5.900	0.7715	1536	0.7259	1083	.9234	1.1385	28.33
8800	5.864	0.7667	1524	0.7203	1083	.9234	1.1409	27.62
9000	5.827	0.7620	1513	0.7148	1082	.9242	1.1456	26.91
9200	5.791	0.7572	1501	0.7093	1081	.9251	1.1492	26.19
9400	5.755	0.7525	1489	0.7039	1080	.9259	1.1528	25.48
9600	5.719	0.7478	1478	0.6984	1079	.9268	1.1564	24.77
9800	5.683	0.7431	1467	0.6931	1079	.9268	1.1600	24.05
10000	5.648	0.7385	1455	0.6877	1078	.9276	1.1637	23.34
10500	5.559	0.7269	1427	0.6745	1076	.9294	1.1729	21.56
11000	5.472	0.7156	1400	0.6614	1074	.9311	1.1822	19.77
11500	5.386	0.7043	1372	0.6486	1072	.9328	1.1916	17.99
12000	5.301	0.6932	1346	0.6360	1070	.9346	1.2011	16.21
12500	5.217	0.6822	1320	0.6236	1068	.9363	1.2107	14.42
13000	5.134	0.6713	1294	0.6113	1066	.9381	1.2205	12.64
13500	5.052	0.6061	1268	0.5993	1064	.9398	1.2304	10.86
14000	4.971	0.6500	1243	0.5874	1062	.9416	1.2403	9.07
14500	4.891	0.6396	1218	0.5758	1060	.9434	1.2504	7.29
15000	4.812	0.6292	1194	0.5643	1058	.9452	1.2606	5.51
15500	4.734	0.6190	1170	0.5531	1056	.9470	1.2710	3.73
16000	4.657	0.6090	1147	0.5420	1054	.9488	1.2814	1.94
16500	4.581	0.5990	1124	0.5411	1052	.9506	1.2921	.16
17000	4.506	0.5892	1101	0.5203	1050	.9524	1.3028	-1.63
17500	4.432	0.5795	1079	0.5098	1048	.9542	1.3137	-3.48
18000	4.358	0.5699	1057	0.4994	1046	.9560	1.3246	-5.19
18500	4.286	0.5604	1035	0.4892	1043	.9588	1.3358	-6.97
19000	4.215	0.5512	1014	0.4791	1041	.9606	1.3470	-8.76
19500	4.144	0.5419	993	0.4693	1029	.9625	1.3584	-10.54
20000	4.075	0.5328	972	0.4595	1037	.9643	1.3700	-12.32
22000	3.805	0.4976	894	0.4223	1029	.9718	1.4176	-19.46
24000	3.550	0.4642	820	0.3876	1021	.9794	1.4678	-26.59
28000	3.078	0.4025	688	0.3250	1004	.9960	1.5762	-40.85
30000	2.861	0.3741	628	0.2970	995	1.0050	1.6349	-47.99

C. BALLISTIC DRAG COEFFICIENT FUNCTIONS FOR VARIOUS BOMBS

Both numerical data and graphs are given to show the ballistic drag coefficient (K_D) functions for different bombs. By reference to Table 10, it may be noted that in several cases many bombs can be represented by the the same K_D curve; the value of c, the reciprocal ballistic coefficient, may be adjusted to match the various K_D curves if two bombs have the same general shape for their representative curves (see eq. 5).

In Table 6, K_D refers only to the ballistic drag coefficient. For comparison with other tables using the aerodynamic drag coefficient C_D, it should be recalled that:

$$K_D = \frac{\pi}{8} C_D$$

The various bombs are denoted by numbers; the numbering scheme is as follows:

Bomb No.	Description of Bomb or Type of K_D Curve
1	Standard G_1 drag function
2	Mk 83/2&3/E
3	Mk 76/0&2/N
4	Mk 76/4/T/L
5	Mk 76/4/T/N
6	HD-200 (fictitious)
7	Mk 43/0 (Nose Mk 43/1)/large fin
8	AN-M57A1 M126 fin
9	AN-M64A1 M128A1 fin

See paragraph F (page 85) for explanation of bomb designation abbreviations.

Figure 19 provides in graphical form the same information as Table 6.

NOTS TP 3902

TABLE 6. Ballistic Drag Coefficient Functions

Mach no.	Bomb no.								
	1	2	3	4	5	6	7	8	9
.00	-	-	-	.095	.079	.1066	.0647	.077	.067
.05	.1003	.0428	.0828	.095	.079	.1066	.0647	.077	.067
.10	.0975	.0428	.0805	.095	.079	.1066	.0647	.077	.067
.15	.0948	.0428	.0783	.095	.079	.1066	.0647	.077	.067
.20	.0921	.0428	.0761	.095	.079	.1066	.0647	.077	.067
.25	.0895	.0428	.0739	.095	.079	.1066	.0647	.077	.067
.30	.0869	.0428	.0718	.095	.079	.1066	.0647	.077	.067
.35	.0846	.0428	.0699	.095	.079	.1066	.0647	.077	.067
.40	.0826	.0428	.0682	.095	.079	.1066	.0647	.077	.067
.45	.0810	.0428	.0669	.095	.079	.1066	.0647	.077	.067
.50	.0799	.0428	.0660	.095	.079	.1066	.0647	.077	.067
.55	.0794	.0428	.0656	.097	.079	.1066	.0647	.077	.067
.60	.0799	.0428	.0660	.098	.080	.1066	.0647	.077	.167
.65	.0816	.0428	.0674	.0995	.0815	.1066	.0647	.077	.067
.70	.0850	.0428	.0702	.1025	.083	.1066	.0647	.077	.067
.75	.0908	.0428	.0750	.107	.0845	.1066	.0647	.077	.067
.80	.1000	.0428	.0826	.113	.087		.0647	.080	.069
.82	.1048	.0429	.0866	.117	.089		.0647	.081	.072
.84	.1106	.0434	.0914	.121	.091		.0650	.085	.074
.86	.1174	.0451	.0970	.126	.095		.0653	.089	.078
.88	.1251	.0479	.1033	.132	.100		.0658	.094	.085
.90	.1340	.0527	.1107	.140	.106		.0662	.106	.094
.91	-	-	-	.145	.110		-	.117	.104
.92	.1438	.0595	.1188	.151	.114		.0668	.131	.118
.93	-	-	-	.157	.119		-	.143	.135
.94	.1546	.0671	.1277	.165	.124		.0680	.162	.157
.95	-	-	-	.174	.130		-	.176	.175
.96	.1660	.0756	.1371	.183	.137		.0701	.193	.193
.97	-	-	-	.194	.144		-	.205	.211
.98	.1774	.0852	.1465	.205	.153		.0755	.220	.227
.99	-	-	-	.217	.163		-	.231	.242
1.00	.1885	.0958	.1557	.229	.175		.0867	.240	.254
1.01	-	-	-	.240	.204		-	.245	.263
1.02	-	.1074	-	.251	.224		.0985	.249	.270
1.05	.2130	.1238	.1759	.277	.248		.1067	.255	.282
1.10	.2308	.1307	.1906	.299	.270		.1100	.262	.290
1.15	.2430	.1326	.2007	.313	.285		.1112	.267	.295
1.25	.2559	-	.2114	.333	.308		.1112	.272	.301
1.50	-	.1387	-	.356	.332		.1067	.277	.306

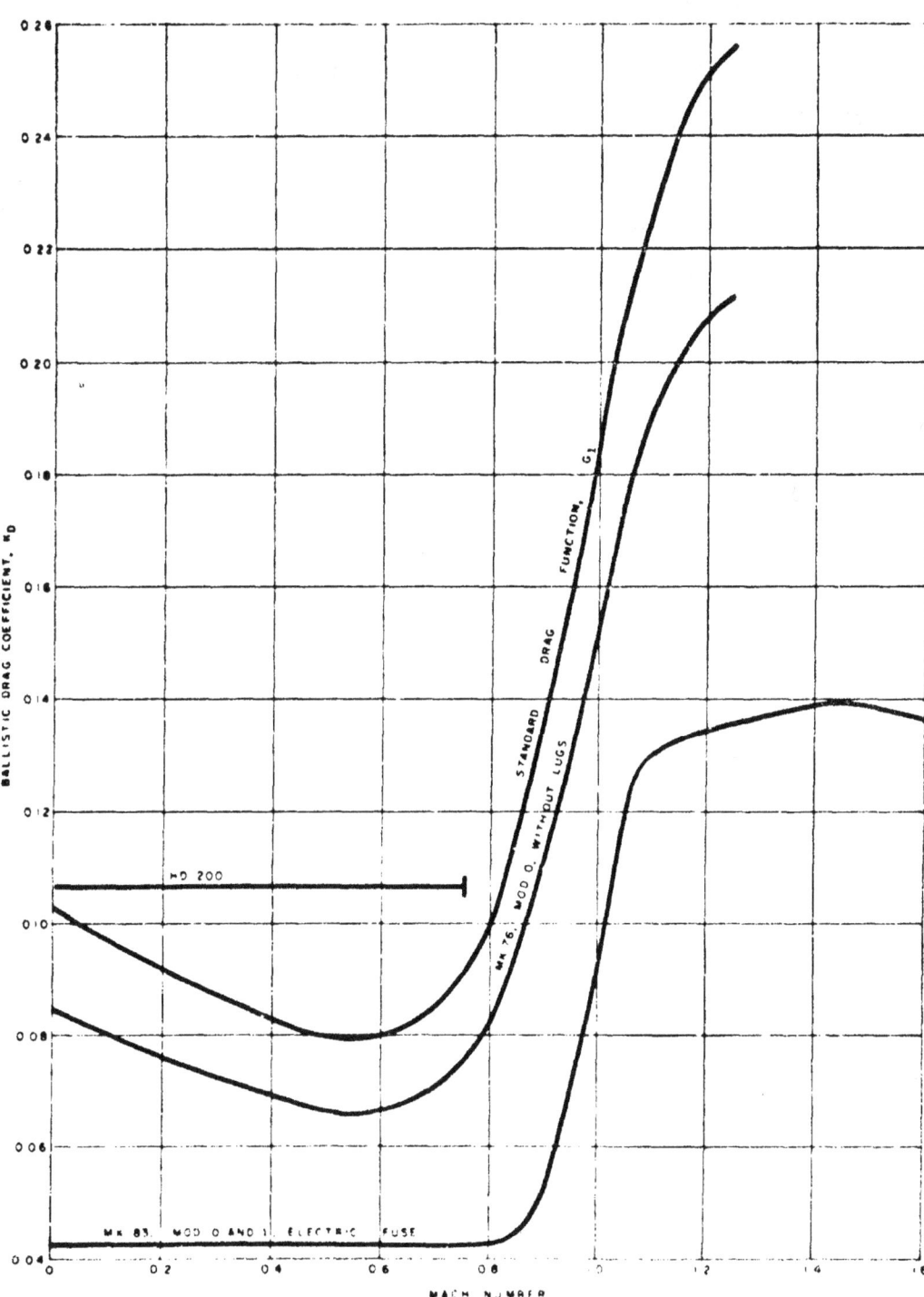

FIG. 19. Ballistic Drag Coefficient Functions.

NOTS TP 3902

D. BALLISTIC TABLES FOR THE MK 83, MK 76, AND HD-200 BOMBS

On the following pages, numerical data are given for the Mk 83/2&3/E (Table 7), the Mk 76/0&2/N (Table 8), and the fictitious retarded bomb, the HD-200 (Table 9).[4] The following parameters are computed for families of release velocities, angles, and altitudes:

- X Ground range, feet
- R Slant range ($R^2 = X^2 + Z^2$), feet
- ϕ Line of sight or harp angle, degrees
- γ Ballistic lead angle, degrees
- t_f Time of flight, seconds
- ψ_e Value of ψ computed by working from computed parameters (the exact value of ψ)
- $\ln \psi_e$ Logarithm of ψ_e
- $\ln \psi_c$ $2/3\, cK_D\, X \sec \theta$ (to be compared with the exact value of $\ln \psi_e$)
- τ_i Impact angle, degrees
- U_i Impact velocity, feet per second
- ΔX Change in ground range caused by change of one degree in release angle, feet

---

[4] See page 85 for explanation of bomb designation notation and bomb data.

TABLE 7. Ballistic Data for the Mk 83 Mod 2&3/L

Z ft	X ft	R ft	φ deg	γ deg	t_f sec	ψ_e	$\ln\psi_e$	$\ln\psi_c$	r_i deg	U_i fps	ΔX ft
\multicolumn{12}{c}{U_o = 400 fps θ = -40 deg}
5000	3468	6084.9	55.25	15.25	11.447	1.014	.01410	.01174	64.05	632	80
10000	5529	11426.7	61.06	21.06	18.395	1.023	.02322	.01604	70.40	873	115
\multicolumn{12}{c}{U_o = 400 fps θ = -20 deg}
1000	1766	2029.5	29.52	9.52	4.719	1.005	.00598	.00549	37.59	470	52
2000	2879	3505.5	34.79	14.79	7.714	1.008	.00886	.00870	45.83	531	67
3000	3761	4810.9	38.58	18.58	10.104	1.012	.01262	.01192	51.05	586	78
4000	4516	6032.8	41.53	21.53	12.155	1.014	.01459	.01285	54.78	635	84
5000	5184	7202.4	43.96	23.96	13.980	1.017	.01725	.01431	57.63	681	91
10000	7826	12698.3	51.95	31.95	21.255	1.025	.02513	.01851	65.77	871	100
\multicolumn{12}{c}{U_o = 400 fps θ = -10 deg}
1000	2363	2565.9	22.94	12.94	6.033	1.007	.00767	.00702	33.89	469	70
2000	3607	4124.4	29.01	19.01	9.236	1.011	.01124	.01039	43.14	530	79
3000	4569	5465.9	32.29	22.29	11.726	1.013	.01380	.01278	48.86	585	84
4000	5381	6704.9	36.63	26.63	13.838	1.016	.01637	.01461	52.88	634	87
5000	6096	7884.2	39.36	29.36	15.706	1.018	.01872	.01605	55.92	680	91
10000	8903	13388.9	48.32	38.42	23.088	1.025	.02557	.02009	64.57	870	101
\multicolumn{12}{c}{U_o = 400 fps θ = 0 deg}
1000	3139	3294.4	17.67	17.67	7.905	1.009	.00936	.00918	32.63	468	86
2000	4431	4861.5	24.29	24.29	11.191	1.013	.01361	.01257	42.28	529	85
3000	5419	6194.0	28.97	28.97	13.717	1.016	.01597	.01493	48.17	584	84
4000	6250	7420.4	32.62	32.62	15.852	1.018	.01833	.01671	52.30	633	84
5000	6981	8586.9	35.61	35.61	17.737	1.020	.02019	.01811	55.42	679	84
10000	9838	14028.1	45.47	45.47	25.157	1.027	.02723	.02187	64.22	869	83
\multicolumn{12}{c}{U_o = 400 fps θ = 20 deg}
1000	4913	5013.7	11.51	31.51	13.216	1.014	.01440	.01528	37.95	466	77
2000	6012	6335.9	18.40	38.40	16.213	1.017	.01755	.01816	46.20	527	62
3000	6885	7510.2	23.54	43.54	18.602	1.020	.01990	.02013	51.41	582	51
4000	7632	8616.7	27.66	47.66	20.653	1.021	.02176	.02171	55.12	632	43
5000	8295	9685.4	31.08	51.08	22.482	1.023	.02333	.02290	57.95	678	37
10000	10923	14800.2	42.47	62.47	29.755	1.028	.02830	.02584	65.99	869	15

TABLE 7. (Cont'd)

Z ft	X ft	R ft	ϕ deg.	γ deg.	t_f sec	ψ_e	$\ln\psi_e$	$\ln\psi_c$	τ_i deg.	U_i fps	ΔX ft
\multicolumn{12}{c}{U_o = 600 fps $\quad \theta$ = -40 deg}
5000	4274	6577.8	49.48	9.48	9.418	1.016	.01617	.01447	56.43	809	113
10000	7152	12294.4	54.43	14.43	15.980	1.026	.02625	.02075	63.15	969	168
\multicolumn{12}{c}{U_o = 600 fps $\quad \theta$ = -20 deg}
1000	2119	2343.1	25.26	5.26	3.778	1.006	.00668	.00659	30.14	645	79
2000	3636	4149.8	28.81	8.81	6.509	1.011	.01124	.01098	36.45	688	111
3000	4880	5728.4	31.58	11.58	8.758	1.015	.01538	.01430	41.01	729	131
4000	5959	7177.0	33.87	13.87	10.723	1.018	.01882	.01695	44.56	767	145
5000	6925	8541.4	35.83	15.83	12.489	1.021	.02147	.01911	47.43	804	157
10000	10787	14709.2	42.83	22.83	19.615	1.031	.03101	.02551	56.52	964	194
\multicolumn{12}{c}{U_o = 600 fps $\quad \theta$ = -10 deg}
1000	3114	4370.6	17.80	7.80	5.307	1.009	.00916	.00924	25.05	643	129
2000	4921	5311.9	22.12	12.12	8.422	1.014	.01469	.01418	32.63	685	151
3000	6333	7007.6	25.35	15.35	10.874	1.019	.01892	.01771	37.85	726	164
4000	7532	8528.3	27.97	17.97	12.965	1.022	.02196	.02045	41.81	764	171
5000	8590	9939.2	30.20	20.20	14.821	1.025	.02489	.02262	44.98	801	178
10000	12753	16206.1	38.10	28.10	22.178	1.034	.03382	.02878	54.80	962	197
\multicolumn{12}{c}{U_o = 600 fps $\quad \theta$ = 0 deg}
1000	4698	4803.3	12.02	12.02	7.913	1.013	.01380	.01373	23.21	639	194
2000	6626	6921.3	16.80	16.80	11.207	1.019	.01921	.01880	31.38	682	192
3000	8099	8636.8	20.33	20.33	13.741	1.023	.02323	.02231	36.88	722	189
4000	9337	10157.7	23.19	23.19	15.881	1.026	.02645	.02496	41.00	761	189
5000	10426	11562.9	25.62	25.62	17.773	1.029	.02898	.02704	44.27	798	188
10000	14684	17765.7	34.26	34.26	25.215	1.037	.03720	.03264	54.34	959	185
\multicolumn{12}{c}{U_o = 600 fps $\quad \theta$ = 20 deg}
1000	9122	9165.7	6.26	26.26	16.492	1.027	.02713	.02835	30.82	630	193
2000	10593	10780.2	10.69	30.69	19.280	1.031	.03072	.03200	37.21	675	175
3000	11810	12185.1	14.25	34.25	21.486	1.034	.03343	.03462	41.80	717	154
4000	12868	13475.0	17.27	37.27	23.453	1.036	.03566	.03661	45.33	756	140
5000	13818	14694.8	19.89	39.89	25.224	1.037	.03730	.03814	48.19	794	127
10000	17634	20272.1	29.56	49.56	32.433	1.043	.04239	.04171	57.12	958	88

TABLE 7. (Cont'd)

Z ft	X ft	R ft	ϕ deg	γ deg	t_f sec	Ψ_e	$\ln\Psi_e$	$\ln\Psi_c$	r_i deg	U_i fps	ΔX ft
\multicolumn{12}{c}{U_o = 800 fps θ = -40 deg}											
5000	4774	6913.1	46.32	6.32	7.898	1.018	.01823	.01616	51.55	957	138
10000	8297	12993.9	50.32	10.32	13.879	1.030	.02985	.02409	57.78	1079	214
\multicolumn{12}{c}{U_o = 800 pfs θ = -20 deg}											
1000	2323	2529.1	23.29	3.29	3.107	1.006	.00578	.00723	26.46	830	99
2000	4138	4596.0	25.80	5.80	5.560	1.013	.01321	.01250	31.14	861	150
3000	5677	6420.9	27.85	7.85	7.654	1.017	.01764	.01664	34.84	891	182
4000	7035	8092.7	29.62	9.62	9.513	1.021	.02157	.02001	37.88	920	206
5000	8263	9658.0	31.18	11.18	11.205	1.025	.02489	.02281	40.46	948	225
10000	13241	16592.9	37.06	17.06	18.161	1.038	.03739	.03132	49.38	1069	286
\multicolumn{12}{c}{U_o = 800 pfs θ = -10 deg}											
1000	3670	3803.8	15.24	5.24	4.697	1.010	.01084	.01089	20.29	826	191
2000	5984	6309.4	18.48	8.48	7.699	1.018	.01794	.01724	26.34	855	225
3000	7819	8374.8	20.99	10.99	10.099	1.023	.02293	.02187	30.80	885	257
4000	9385	10201.9	23.08	13.08	12.161	1.027	.02693	.02548	34.35	914	272
5000	10772	11875.9	24.90	14.90	13.998	1.031	.03053	.02837	37.30	942	283
10000	16245	19076.2	31.62	21.62	21.340	1.043	.04239	.03667	47.17	1064	316
\multicolumn{12}{c}{U_o = 800 pfs θ = 0 deg}											
1000	6250	6329.5	9.90	9.90	7.923	1.018	.01833	.01827	17.90	817	346
2000	8808	9032.2	12.79	12.79	11.225	1.025	.02528	.02500	24.71	846	339
3000	10759	11169.4	15.58	15.58	13.765	1.031	.03063	.02964	29.54	876	335
4000	12399	13028.2	17.88	17.88	15.913	1.035	.03450	.03315	33.32	906	332
5000	13841	14716.4	19.86	19.86	17.811	1.038	.03758	.03590	36.41	934	331
10000	19473	21890.6	27.18	27.18	25.301	1.049	.04803	.04328	46.62	1059	325
\multicolumn{12}{c}{U_o = 800 pfs θ = 20 deg}											
1000	14560	14594.3	3.93	23.93	20.002	1.043	.04296	.04530	27.48	794	396
2000	16312	16434.2	6.99	26.99	22.476	1.047	.04679	.04927	32.32	828	345
3000	17796	18047.1	9.57	29.57	24.579	1.051	.05003	.05218	36.08	860	313
4000	19113	19527.1	11.82	31.82	26.448	1.053	.05221	.05438	39.15	892	289
5000	20312	20918.3	13.83	33.83	28.150	1.055	.05373	.05607	41.73	923	269
10000	25200	27111.6	21.64	41.64	35.110	1.060	.05883	.05961	50.46	1055	207

TABLE 7. (Cont'd)

Z ft	X ft	R ft	ϕ deg	γ deg	t_f sec	Ψ_e	$\ln\Psi_e$	$\ln\Psi_c$	r_i deg	U_i fps	ΔX ft
\multicolumn{12}{c}{U_o = 1000 fps θ = -40 deg}
5000	5089	7134.3	44.49	4.49	6.794	1.028	.02761	.02259	48.52	1099	155
10000	9070	13500.6	47.79	7.79	12.426	1.060	.05779	.03892	54.22	1164	250
\multicolumn{12}{c}{U_o = 1000 fps θ = -20 deg}
1000	2446	2642.5	22.24	2.24	2.622	1.007	.00668	.00937	24.44	1017	113
2000	4475	4901.6	24.08	4.08	4.829	1.018	.01745	.01664	28.00	1034	180
3000	6241	6924.6	25.67	5.67	6.776	1.027	.02625	.02338	30.99	1050	227
4000	7823	8786.3	27.08	7.08	8.543	1.034	.03324	.02902	33.59	1066	261
5000	9265	10528.1	28.35	8.35	10.176	1.041	.04009	.03352	35.89	1081	288
10000	15145	18148.6	33.44	3.44	17.095	1.074	.07130	.05298	44.59	1143	373
\multicolumn{12}{c}{U_o = 1000 fps θ = -10 deg}
1000	4081	4201.7	13.77	3.77	4.191	1.015	.01499	.01491	17.48	1009	249
2000	6834	7120.6	16.31	6.31	7.076	1.026	.02586	.02425	22.37	1023	320
3000	9046	9530.5	18.35	8.35	9.431	1.035	.03450	.03234	26.23	1038	357
4000	10941	11649.3	20.08	10.08	11.477	1.043	.04201	.03872	29.42	1054	380
5000	12620	13574.4	21.61	11.61	13.316	1.050	.04917	.04357	32.16	1069	397
10000	19196	21644.5	27.52	17.52	20.797	1.082	.07909	.06408	42.07	1132	443
\multicolumn{12}{c}{U_o = 1000 fps θ = 0 deg}
1000	7778	7842.0	7.33	7.33	7.940	1.028	.02713	.02800	14.62	992	536
2000	10940	11121.3	10.36	10.36	11.256	1.039	.03807	.03823	20.45	1007	523
3000	13343	13676.1	12.67	12.67	13.819	1.048	.04641	.04698	24.74	1023	515
4000	15355	15867.5	14.60	14.60	15.991	1.055	.05316	.05352	28.20	1040	510
5000	17115	17830.4	16.29	16.29	17.918	1.061	.05931	.05819	31.13	1056	507
10000	23889	25897.6	22.71	22.71	25.608	1.089	.08554	.07854	41.51	1125	503
\multicolumn{12}{c}{U_o = 1000 fps θ = 20 deg}
1000	21122	21145.7	2.71	22.71	23.602	1.069	.06663	.08092	25.88	948	628
2000	23049	23135.6	4.96	24.96	25.840	1.073	.07083	.08573	29.63	974	566
3000	24729	24910.3	6.92	26.92	27.797	1.077	.07437	.09267	32.71	998	523
4000	26242	26545.1	8.67	28.67	29.568	1.080	.07715	.09735	35.34	1021	489
5000	27630	28078.8	10.26	30.26	31.202	1.083	.07937	.09998	37.63	1043	462
10000	33350	34817.0	16.69	36.69	38.075	1.093	.08856	.11660	46.06	1126	381

TABLE 9. Ballistic Data for the Mk 76 Mod 0&2/N

Z ft	X ft	R ft	ϕ deg	γ deg	t_f sec	ψ_e	$\ln\psi_e$	$\ln\psi_c$	γ_i deg	U_i fps	ΔX ft
\multicolumn{12}{c}{U_o = 400 fps θ = -40 deg}
5000	3404	6048.7	55.75	15.75	11.303	1.080	.07678	.06366	65.14	637	78
10000	5345	11338.8	61.08	21.88	19.231	1.127	.11929	.08541	72.00	781	110
\multicolumn{12}{c}{U_o = 400 fps θ = -20 deg}
1000	1754	2019.0	29.69	9.69	4.780	1.032	.03169	.03022	38.07	454	51
2000	2839	3472.7	35.16	15.16	7.859	1.053	.05202	.04749	46.78	505	66
3000	3689	4754.9	39.12	19.12	10.337	1.070	.06728	.05982	52.36	551	74
4000	4407	5951.6	42.23	22.23	12.480	1.084	.08020	.06925	56.36	592	70
5000	5038	7098.0	44.78	24.78	14.401	1.096	.09130	.07681	59.41	629	85
10000	7488	12492.8	53.17	33.17	22.174	1.139	.13059	.09754	68.11	755	102
\multicolumn{12}{c}{U_o = 400 fps θ = -10 deg}
1000	2335	2540.1	23.18	13.18	6.110	1.041	.03989	.03839	34.64	450	67
2000	3535	4061.6	29.50	19.50	9.401	1.063	.06081	.05643	44.46	500	74
3000	4450	5366.8	33.99	23.99	11.981	1.079	.07613	.06885	50.56	546	78
4000	5214	6571.6	37.49	27.49	14.186	1.093	.08902	.07818	54.85	588	81
5000	5882	7720.0	40.37	30.37	16.149	1.105	.09975	.08557	58.10	625	83
10000	8461	13099.2	49.77	39.77	24.027	1.146	.13663	.10510	67.25	772	91
\multicolumn{12}{c}{U_o = 400 fps θ = 0 deg}
1000	3074	3232.6	18.02	18.02	7.990	1.053	.05117	.04978	33.76	444	81
2000	4303	4745.1	24.93	24.93	11.365	1.074	.07167	.06765	44.03	495	78
3000	5230	6029.3	29.84	29.84	13.980	1.091	.08691	.07969	50.30	542	77
4000	6004	7214.4	33.67	33.67	16.207	1.105	.09966	.08865	54.67	584	75
5000	6678	8342.4	36.82	36.82	18.187	1.115	.10894	.09567	57.97	622	75
10000	9285	13645.9	47.12	47.12	26.101	1.154	.14297	.11360	67.19	770	72
\multicolumn{12}{c}{U_o = 400 fps θ = 20 deg}
1000	4683	4788.6	12.05	32.05	13.230	1.083	.07983	.08070	40.02	434	68
2000	5704	6044.5	19.32	39.32	16.319	1.100	.09558	.09543	48.77	489	51
3000	6506	7164.4	24.76	44.76	18.799	1.114	.10769	.10550	54.23	537	42
4000	7189	8226.9	29.09	49.09	20.944	1.124	.11725	.11290	58.08	580	35
5000	7793	9259.1	32.68	52.68	22.868	1.133	.12513	.11880	61.00	619	29
10000	10173	14265.0	44.51	64.51	30.638	1.163	.15092	.13250	69.16	770	8

TABLE 8. (Cont'd)

Z ft	X ft	R ft	ϕ deg	γ deg	t_f sec	ψ_e	$\ln\psi_e$	$\ln\psi_c$	γ_i deg	U_i fps	ΔX ft
\multicolumn{12}{c}{$U_o = 600$ fps $\theta = -40$ deg}
5000	4208	6535.0	49.92	9.92	9.788	1.090	.08572	.07439	57.55	741	109
10000	6922	12162.0	55.31	15.31	16.839	1.149	.13880	.10500	65.21	842	159
\multicolumn{12}{c}{$U_o = 600$ fps $\theta = -20$ deg}
1000	2018	2333.2	25.38	5.38	3.840	1.035	.03440	.03429	30.50	620	78
2000	3591	4110.4	29.12	9.12	6.668	1.062	.06006	.05671	37.33	645	107
3000	4768	5650.2	32.07	12.07	9.028	1.084	.08047	.07328	42.38	671	124
4000	5814	7057.1	34.53	14.53	11.105	1.101	.09649	.08622	46.34	697	135
5000	6721	8376.9	36.65	16.65	12.988	1.117	.11082	.09686	49.57	722	145
10000	10263	14329.3	44.26	24.26	20.722	1.175	.16152	.12690	59.83	826	173
\multicolumn{12}{c}{$U_o = 600$ fps $\theta = -10$ deg}
1000	3075	3233.5	18.01	8.01	5.401	1.051	.04955	.04773	25.72	608	124
2000	4811	5210.2	22.57	12.57	8.629	1.080	.07687	.07250	34.02	632	141
3000	6143	6836.4	26.03	16.03	11.196	1.102	.09749	.08972	39.83	659	150
4000	7258	4905.0	28.86	18.86	13.402	1.121	.11395	.10270	44.27	685	154
5000	8232	9631.5	31.27	21.27	15.373	1.137	.12795	.11320	47.82	711	158
10000	11987	15610.5	39.84	29.84	23.333	1.191	.17496	.14140	58.79	818	169
\multicolumn{12}{c}{$U_o = 600$ fps $\theta = 0$ deg}
1000	4564	4672.3	12.36	12.36	8.028	1.074	.07167	.06976	24.45	591	178
2000	6367	6673.7	17.44	17.44	11.435	1.104	.09903	.09449	33.51	617	169
3000	7720	8282.4	21.24	21.24	14.801	1.127	.11912	.11100	39.65	645	165
4000	8846	9708.3	24.33	24.33	16.334	1.144	.13444	.12320	44.25	673	161
5000	9828	11026.8	26.96	26.96	18.339	1.158	.14704	.13300	47.89	700	158
10000	13613	16891.2	36.30	36.30	26.370	1.208	.18863	.15820	58.94	813	151
\multicolumn{12}{c}{$U_o = 600$ fps $\theta = 20$ deg}
1000	8354	8413.6	6.83	26.83	16.373	1.144	.13462	.13590	33.89	555	166
2000	9680	9834.5	11.67	31.67	19.238	1.165	.15255	.15280	41.07	591	135
3000	10759	11169.4	15.58	35.58	21.613	1.181	.16602	.16460	46.14	626	116
4000	11963	12358.2	18.89	38.89	23.698	1.193	.17664	.17340	49.99	659	103
5000	12530	13490.8	21.75	41.75	25.585	1.203	.18507	.18050	53.05	689	92
10000	15391	18775.6	32.18	52.18	33.311	1.235	.21115	.19650	62.37	811	59

TABLE 8. (Cont'd)

Z ft	X ft	R ft	ϕ deg	γ deg	t_f sec	ψ_e	$\ln\psi_e$	$\ln\psi_c$	r_i deg	U_i fps	ΔX ft
\multicolumn{12}{c}{U_o = 800 fps   = -40 deg}
5000	4704	6865.0	46.75	6.75	8.322	1.111	.10517	.09266	52.73	852	132
10000	8002	12807.5	51.33	11.33	15.072	1.198	.18057	.13690	60.41	894	199
\multicolumn{12}{c}{U_o = 800 fps   = -20 deg}
1000	2312	2519.0	23.39	3.39	3.174	1.042	.04085	.04121	26.76	790	98
2000	4088	4551.0	26.07	6.07	5.751	1.077	.07371	.07105	31.98	791	144
3000	5564	6321.2	28.33	8.33	7.985	1.106	.10093	.09425	36.26	798	171
4000	6845	7928.1	30.30	10.30	9.995	1.131	.12319	.11230	39.86	807	190
5000	7986	9422.1	32.05	12.05	11.841	1.153	.14245	.12820	42.96	818	204
10000	12452	15970.4	38.77	18.77	19.593	1.239	.21414	.17360	53.83	870	245
\multicolumn{12}{c}{U_o = 800 fps   = -10 deg}
1000	3620	3755.6	15.44	5.44	4.815	1.065	.06288	.06158	20.93	768	178
2000	5827	6160.7	18.94	8.94	7.970	1.105	.10003	.09663	27.82	766	215
3000	7533	8108.4	21.71	11.71	10.525	1.137	.12813	.12170	33.06	773	229
4000	8961	9813.2	24.06	14.06	12.739	1.163	.15083	.14090	37.30	784	237
5000	10207	11365.9	26.10	16.10	14.729	1.185	.16991	.15640	40.83	796	243
10000	14974	18006.1	33.74	23.74	22.833	1.267	.23626	.19930	52.69	858	258
\multicolumn{12}{c}{U_o = 800 fps   = 0 deg}
1000	5998	6080.8	9.47	9.47	8.082	1.106	.10057	.10040	19.33	731	305
2000	8326	8562.8	13.51	13.51	11.532	1.148	.13785	.13590	27.27	733	287
3000	10006	10498.7	16.60	16.60	14.215	1.179	.16475	.16010	33.00	744	276
4000	11498	12173.9	19.18	19.18	16.506	1.204	.18540	.17810	37.50	758	268
5000	12749	13694.4	21.41	21.41	18.548	1.224	.20196	.19230	41.19	774	262
10000	17534	20185.2	29.70	29.70	26.750	1.294	.25774	.22980	53.14	849	249
\multicolumn{12}{c}{U_o = 800 fps   = 20 deg}
1000	12617	12656.6	4.53	24.53	19.558	1.234	.21034	.22490	31.73	654	282
2000	14119	14259.9	8.06	28.06	22.208	1.258	.22960	.24530	37.58	678	241
3000	15381	15670.8	11.04	31.04	24.466	1.277	.24436	.26050	42.02	703	214
4000	16496	16974.0	13.63	33.63	26.486	1.292	.25580	.27190	45.57	728	194
5000	17507	18207.0	15.94	35.94	28.338	1.303	.26498	.28110	48.51	751	179
10000	21648	23846.1	24.79	44.79	36.025	1.340	.29289	.30190	58.05	845	134

TABLE 8 (Cont'd)

Z ft	X ft	R ft	ϕ deg	γ deg	t_f sec	Ψ_e	$\ln\Psi_e$	$\ln\Psi_c$	τ_1 deg	U_1 fps	ΔX ft
\multicolumn{12}{c}{$U_o = 1000$ fps $\quad \theta = -40$ deg}
5000	4998	7069.7	45.01	5.01	7.365	1.177	.16322	.15550	49.96	933	139
10000	8680	13241.7	49.04	9.04	13.912	1.315	.27406	.24610	57.46	930	228
\multicolumn{12}{c}{$U_o = 1000$ fps $\quad \theta = -20$ deg}
1000	2432	2629.6	22.35	2.35	2.717	1.066	.06354	.06642	24.78	940	111
2000	4405	4837.8	24.42	4.42	5.108	1.122	.11529	.11800	29.03	909	171
3000	6076	6776.3	26.28	6.28	7.262	1.172	.15905	.16030	32.80	892	208
4000	7537	8532.7	27.96	7.96	9.240	1.214	.19425	.19500	36.16	924	233
5000	8843	10158.7	29.48	9.48	11.082	1.250	.22346	.22430	39.16	879	251
10000	13926	17144.5	35.68	15.68	18.950	1.396	.33347	.32190	50.44	891	301
\multicolumn{12}{c}{$U_o = 1000$ fps $\quad \theta = -10$ deg}
1000	4003	4126.0	14.03	4.03	4.378	1.107	.10138	.10430	18.30	899	232
2000	6569	6866.7	16.93	6.93	7.499	1.176	.16212	.16800	24.36	865	281
3000	8556	9066.7	19.32	9.32	10.073	1.228	.20555	.21540	29.27	851	300
4000	10211	10966.5	21.39	11.39	12.320	1.272	.24043	.25210	33.39	847	311
5000	11649	12676.7	23.23	13.23	14.354	1.309	.26911	.28200	36.34	848	316
10000	17078	19790.4	30.35	20.35	22.682	1.445	.36783	.37670	49.44	874	331
\multicolumn{12}{c}{$U_o = 1000$ fps $\quad \theta = 0$ deg}
1000	7237	7305.8	7.87	7.87	8.183	1.187	.17134	.18570	16.64	830	434
2000	9953	10152.0	11.36	11.36	11.695	1.255	.22714	.25060	24.00	811	400
3000	11958	12328.6	14.08	14.08	14.428	1.304	.26559	.29650	29.49	808	381
4000	13607	14182.8	16.38	16.38	16.764	1.343	.29491	.33090	33.92	811	367
5000	15032	15841.7	18.40	18.40	18.850	1.376	.31882	.35840	37.65	819	358
10000	20425	22741.6	26.09	26.09	27.261	1.490	.39211	.44360	50.23	864	336
\multicolumn{12}{c}{$U_o = 1000$ fps $\quad \theta = 20$ deg}
1000	16519	16549.2	3.46	23.46	22.289	1.411	.34402	.45110	31.67	714	385
2000	18082	18192.3	6.31	26.31	24.770	1.441	.36513	.48460	36.69	732	338
3000	19433	19663.2	8.78	28.78	26.937	1.464	.38124	.51280	40.66	752	307
4000	20639	21023.0	10.97	30.97	28.900	1.483	.39494	.53410	43.94	771	284
5000	21735	22302.7	12.96	32.96	30.713	1.500	.40553	.55150	46.73	789	266
10000	26335	28169.7	20.79	40.79	38.375	1.550	.43832	.60870	56.14	864	216

TABLE 9. Ballistic Data for the HD-200

Z ft	X ft	R ft	ϕ deg	γ deg	t_f sec	ψ_e	$\ln\psi_e$	$\ln\psi_c$	γ_1 deg	U_1 fps	ΔX ft
\multicolumn{12}{c}{U_o = 400 fps θ = -20 deg}
100	252	271.1	21.64	1.64	.749	1.145	.13540	.1453	23.55	329	12
200	461	502.5	23.45	3.45	1.511	1.331	.28601	.2650	27.73	282	19
300	635	702.3	25.29	5.29	2.265	1.500	.40560	.3640	32.25	251	22
400	780	876.6	27.15	7.15	2.999	1.676	.51641	.4458	36.88	230	24
500	904	1033.1	28.95	8.95	3.711	1.837	.60835	.5152	41.45	216	24
1000	1315	1652.0	37.25	17.25	6.922	2.648	.97380	.7385	59.75	192	22
\multicolumn{12}{c}{U_o = 400 fps θ = -10 deg}
100	429	440.5	13.12	3.12	1.309	1.277	.24412	.2360	17.02	287	27
200	698	726.1	15.99	5.99	2.405	1.523	.42068	.3829	24.28	239	29
300	888	937.3	18.67	8.67	3.357	1.754	.56213	.4857	31.14	214	29
400	1036	1110.5	21.11	11.11	4.215	1.953	.66947	.5650	37.44	200	27
500	1155	1258.6	23.41	13.41	5.008	2.143	.76211	.6281	43.13	192	25
1000	1532	1829.5	33.13	23.13	8.350	3.000	1.09851	.8209	62.50	185	20
\multicolumn{12}{c}{U_o = 400 fps θ = 0 deg}
100	787	793.3	7.24	7.24	2.706	1.606	.47362	.4264	17.91	219	43
200	1024	1043.4	11.05	11.05	4.017	1.897	.64027	.5532	28.29	193	33
300	1183	1220.5	14.23	14.23	5.014	2.132	.75711	.6373	36.51	182	28
400	1303	1363.0	17.07	17.07	5.882	2.343	.85152	.6998	43.30	177	25
500	1400	1486.6	19.65	19.65	6.669	2.537	.93106	.7497	48.99	176	22
1000	1715	1985.3	30.25	30.25	9.959	3.382	1.21835	.9050	66.51	182	15
\multicolumn{12}{c}{U_o = 400 fps θ = 20 deg}
100	1436	1439.5	3.98	23.98	7.415	2.652	.97528	.8280	45.46	155	17
200	1525	1538.1	7.47	27.47	8.266	2.851	1.04781	.8768	52.00	159	14
300	1599	1626.9	10.63	30.63	9.032	3.030	1.10843	.9166	57.04	163	11
400	1661	1708.5	13.54	33.54	9.742	3.198	1.16246	.9494	61.06	167	9
500	1715	1786.4	16.25	36.25	10.410	3.357	1.21102	.9774	64.35	171	7
1000	1910	2156.0	27.63	47.63	13.402	4.081	1.40634	1.072	74.71	185	3

TABLE 9. (Cont'd)

Z ft	X ft	R ft	ϕ deg	γ deg	t_f sec	ψ_e	$\ln\psi_e$	$\ln\psi_c$	τ_i deg	U_i fps	ΔX ft
\multicolumn{12}{c}{U_o = 600 fps θ = -20 deg}
100	263	281.4	20.82	.82	.525	1.222	.20008	.1516	21.69	482	14
200	502	540.4	21.72	1.72	1.116	1.356	.30417	.2886	23.99	398	23
300	713	773.5	22.82	2.82	1.753	1.574	.45349	.4087	26.90	337	30
400	897	982.1	24.03	4.03	2.417	1.806	.59089	.5127	30.31	294	33
500	1057	1169.3	25.32	5.32	3.096	2.039	.71246	.6024	34.17	262	34
1000	1591	1879.2	32.15	12.15	6.334	3.286	1.18967	.8935	53.12	202	32
\multicolumn{12}{c}{U_o = 600 fps θ = -10 deg}
100	487	497.2	11.60	1.60	1.011	1.293	.25689	.2679	13.78	404	38
200	833	856.7	13.50	3.50	2.032	1.662	.50772	.4570	18.94	309	46
300	1085	1125.7	15.46	5.46	2.984	2.004	.69500	.5935	24.64	260	45
400	1277	1338.2	17.39	7.39	3.868	2.327	.84449	.6865	30.54	231	43
500	1429	1514.0	19.28	9.28	4.696	2.636	.96930	.7771	36.37	212	39
1000	1896	2143.6	27.80	17.80	8.166	4.019	1.39103	1.016	57.98	188	28
\multicolumn{12}{c}{U_o = 600 fps θ = 0 deg}
100	1070	1074.7	5.34	5.34	2.853	1.955	.67019	.5798	14.65	256	71
200	1357	1371.7	8.38	8.38	4.158	2.431	.88810	.7331	24.41	213	52
300	1547	1575.8	10.97	10.97	5.189	2.805	1.03147	.8333	32.46	195	42
400	1686	1732.8	13.35	13.35	6.085	3.149	1.14708	.9056	39.73	186	37
500	1794	1862.4	15.57	15.57	6.894	3.477	1.24605	.9607	45.51	181	32
1000	2148	2369.4	24.96	24.96	10.249	4.850	1.57902	1.133	64.44	182	21
\multicolumn{12}{c}{U_o = 600 fps θ = 20 deg}
100	1963	1965.6	2.92	22.92	8.903	4.177	1.42952	1.131	50.80	161	18
200	2040	2049.8	5.60	25.60	9.674	4.475	1.49857	1.172	55.93	164	15
300	2107	2128.3	8.10	28.10	10.388	4.749	1.55789	1.207	60.05	168	9
400	2164	2200.7	10.47	30.47	11.060	5.012	1.61172	1.236	63.44	171	10
500	2215	2270.7	12.72	32.72	11.702	5.261	1.66031	1.262	66.28	175	8
1000	2404	2603.7	22.59	42.59	14.630	6.411	1.8503	1.350	75.56	187	3

TABLE 9. (Cont'd)

Z ft	X ft	R ft	ϕ deg	γ deg	t_f sec	ψ_e	$\ln\psi_e$	$\ln\psi_c$	r_1 deg	U_1 fps	ΔX ft
\multicolumn{12}{c}{U_o = 800 fps $\quad\theta$ = -20 deg}											
100	268	286.0	20.46	.46	.401	1.201	.18340	.1545	20.98	638	15
200	520	557.1	21.04	1.04	.875	1.395	.33268	.2939	22.40	517	27
300	752	809.6	21.75	1.75	1.413	1.634	.49072	.4311	24.34	427	35
400	962	1041.9	22.58	2.58	2.002	1.893	.63800	.5498	26.84	362	40
500	1149	1253.1	23.52	3.52	2.631	2.177	.77776	.6548	29.91	312	43
1000	1786	2046.9	29.24	9.24	5.845	3.854	1.34914	1.003	48.22	213	40
\multicolumn{12}{c}{U_o = 800 fps $\quad\theta$ = -10 deg}											
100	515	524.6	10.99	.99	.813	1.337	.29058	.2833	12.32	524	46
200	917	938.6	12.30	2.30	1.741	1.758	.56406	.5030	16.09	380	60
300	1219	1255.4	13.83	3.83	2.676	2.209	.79236	.6668	20.88	304	60
400	1450	1504.2	15.42	5.42	3.517	2.649	.97403	.7908	26.28	259	56
500	1632	1706.9	17.03	7.03	4.413	3.075	1.12317	.8875	31.81	231	49
1000	2169	2388.4	24.75	14.75	7.994	5.065	1.62230	1.162	54.86	190	35
\multicolumn{12}{c}{U_o = 800 fps $\quad\theta$ = 0 deg}											
100	1307	1310.8	4.38	4.38	2.923	2.329	.84540	.7082	13.02	279	97
200	1628	1640.2	7.00	7.00	4.263	3.002	1.09931	.8796	22.45	225	68
300	1835	1859.4	9.29	9.29	5.318	3.545	1.26540	.9885	30.46	202	54
400	1985	2024.9	11.39	11.39	6.230	4.039	1.39592	1.066	37.52	191	46
500	2100	2158.7	13.39	13.39	7.053	4.511	1.50643	1.124	43.75	184	40
1000	2477	2671.2	21.93	21.98	10.445	6.484	1.86937	1.307	63.40	183	25
\multicolumn{12}{c}{U_o = 800 fps $\quad\theta$ = 20 deg}											
100	2355	2357.1	2.43	22.43	9.954	6.063	1.80218	1.358	54.93	165	18
200	2425	2433.2	4.71	24.71	10.677	6.467	1.86676	1.394	59.20	168	14
300	2485	2503.0	6.83	26.88	11.358	6.852	1.92454	1.424	62.71	171	12
400	2538	2569.3	8.96	28.96	12.006	7.219	1.97678	1.450	65.64	175	10
500	2586	2633.9	10.94	30.94	12.629	7.571	2.02432	1.473	68.14	177	8
1000	2768	2943.1	19.86	39.86	15.509	9.204	2.21968	1.554	76.50	188	2

E. BALLISTIC CURVES AND SENSITIVITY GRAPHS

In this section are given graphs (Fig. 20 through 44) showing the curves of ground range X, time of flight t_f, impact angle τ_i, impact velocity U_i, and lead angle γ, all plotted against release altitude for a family of release velocities, at three release angles, for the Mk 83, Mk 76, and the fictitious HD-200 bombs.

Also included are a number of sensitivity curves for the same bombs and the same release conditions. These curves show the change in the specified trajectory parameter caused by an incremental change in one of the independent release variables Z, U_o, θ, or cK_D. They are of value, for example, in quickly estimating the effects on the trajectory of errors in the release conditions and are useful in various other applications. The figures 400, 600, 800, and 1,000 on the curves refer to release velocity in feet per second.

However, in some applications, the curves must be used rather carefully since they were obtained by varying in a small amount each of the independent release variables Z, θ, U_o, and cK_D, in turn from a given set of "standard" release conditions. Thus, the curves show only the change in the trajectory caused by a slight change in a single release variable. In analyzing many fire control systems, account must also be taken of the specific system mechanization and of the method of aiming before data points of the curves can be applied correctly.

NOTS TP 3902

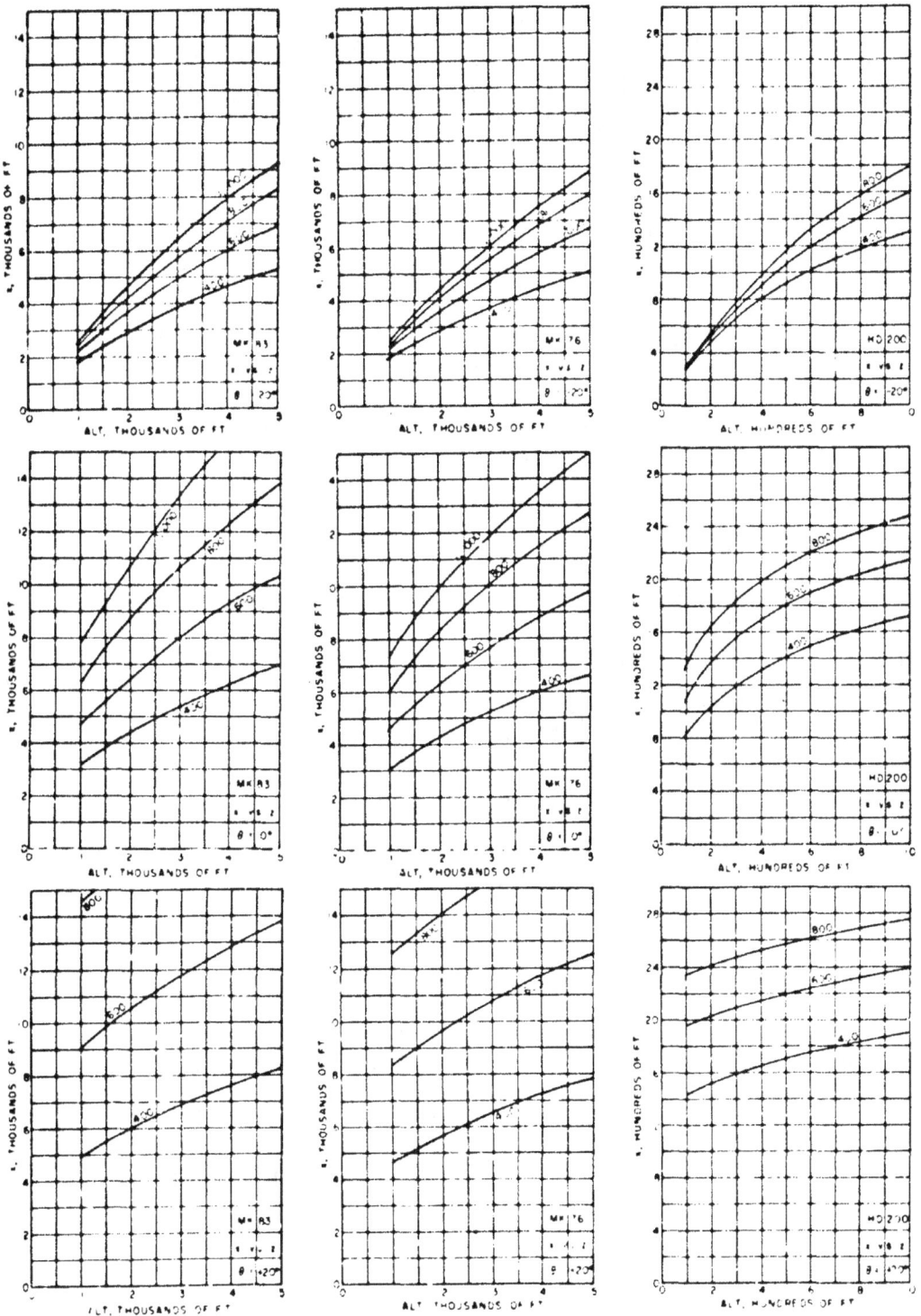

FIG. 20. Ground Range Versus Altitude.

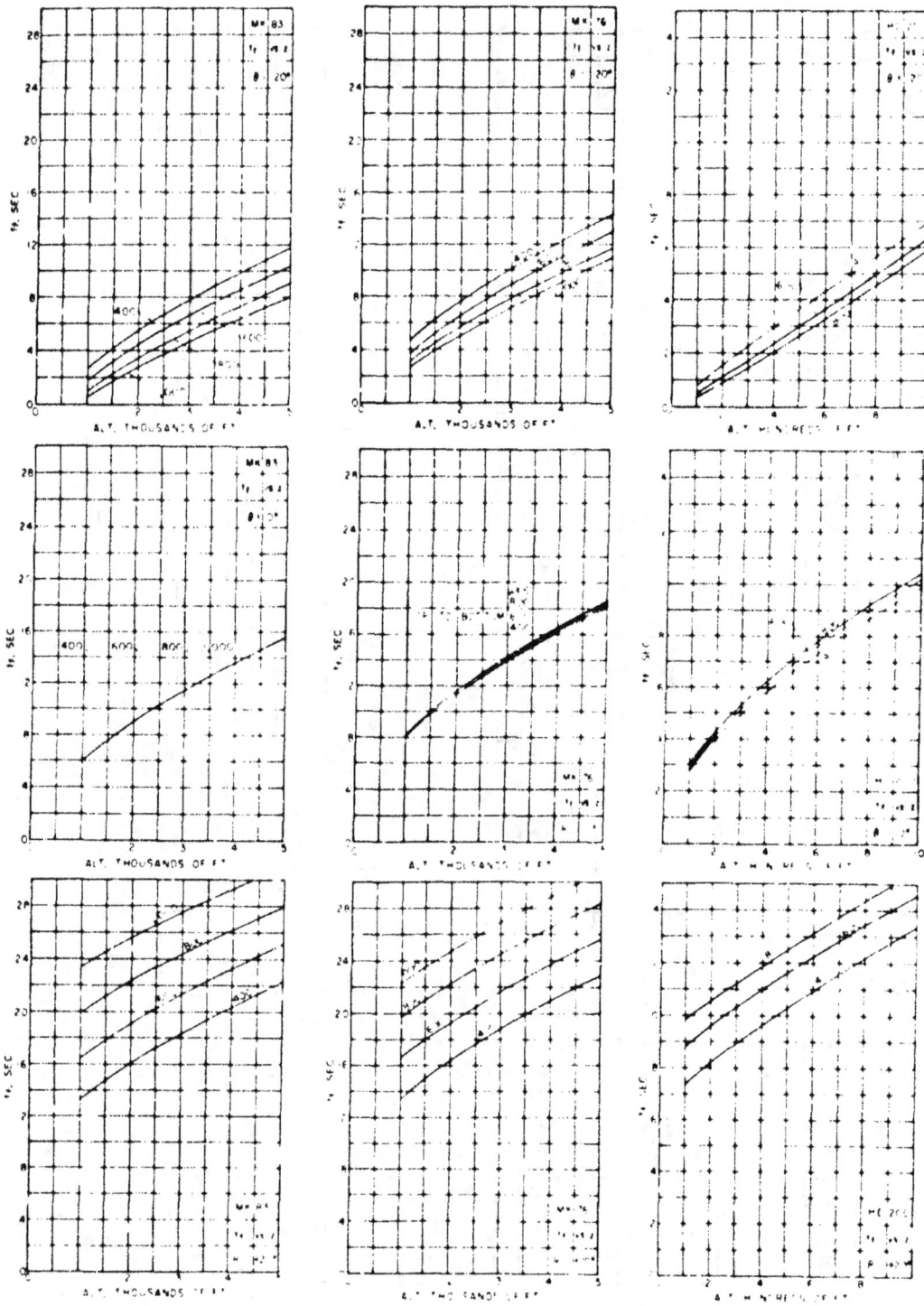

FIG. 21. Time of Flight Versus Altitude.

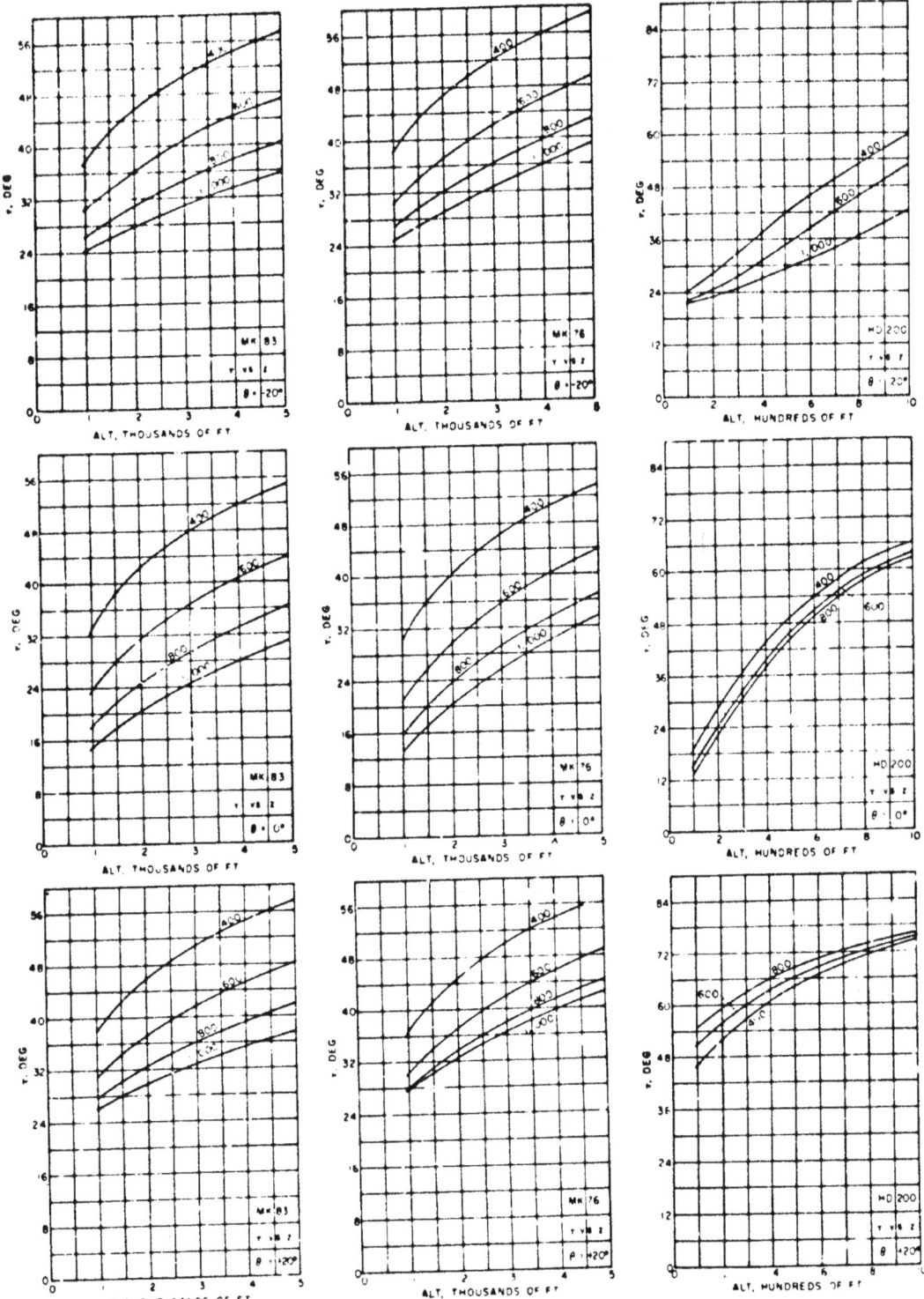

FIG. 22. Impact Angle Versus Altitude.

NOTS TP 3902

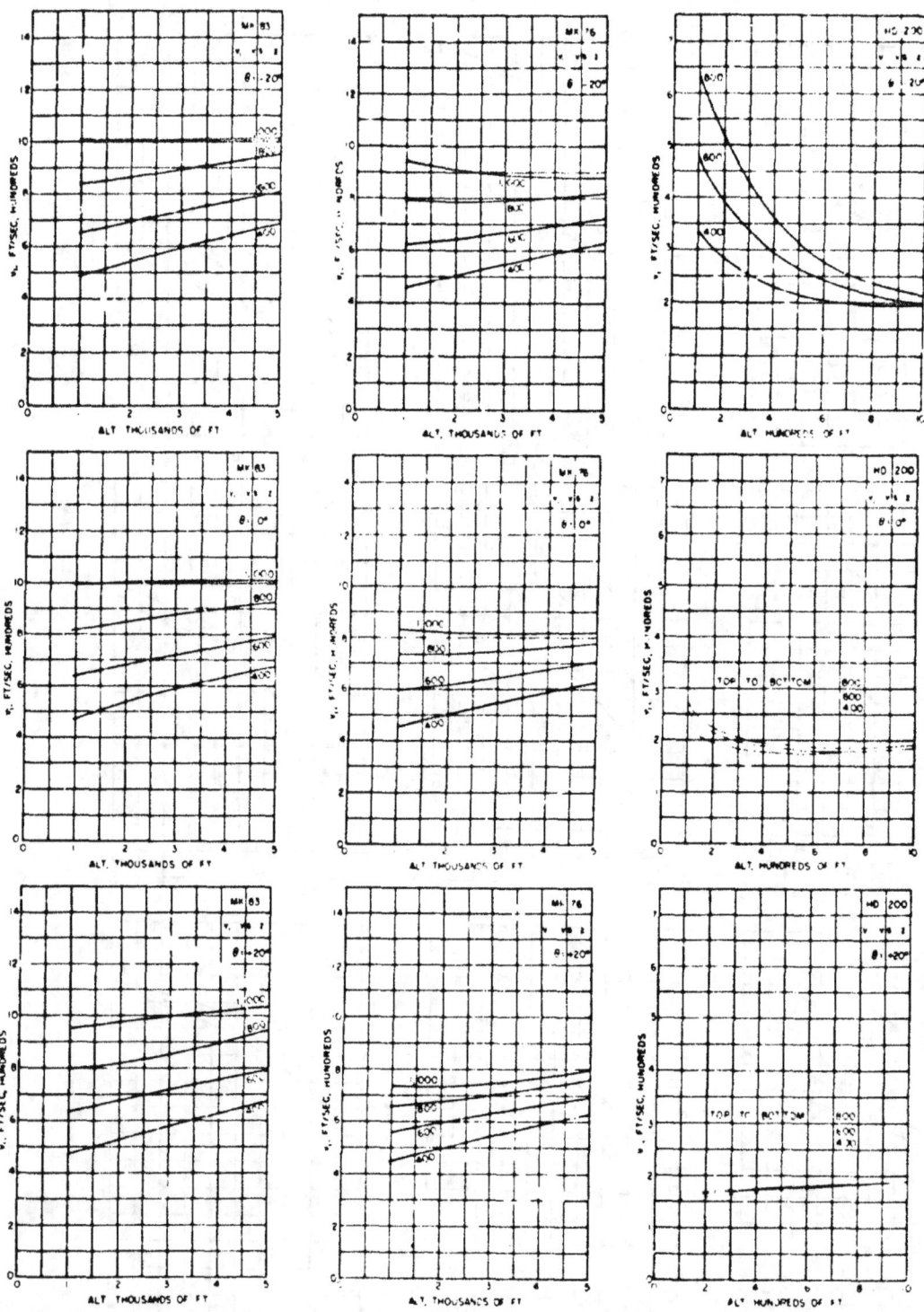

FIG. 23. Impact Velocity Versus Altitude.

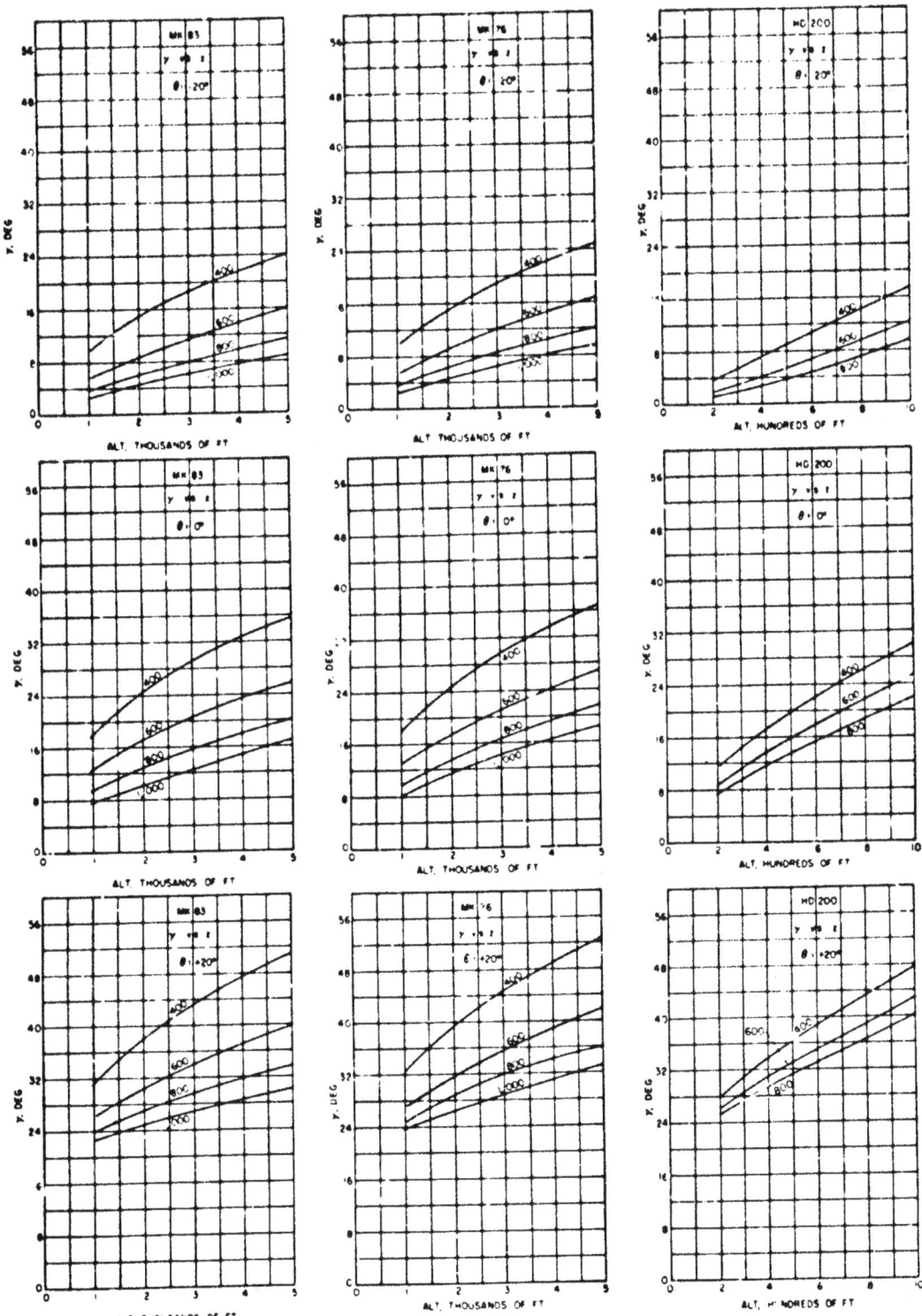

FIG. 24. Lead Angle Versus Altitude.

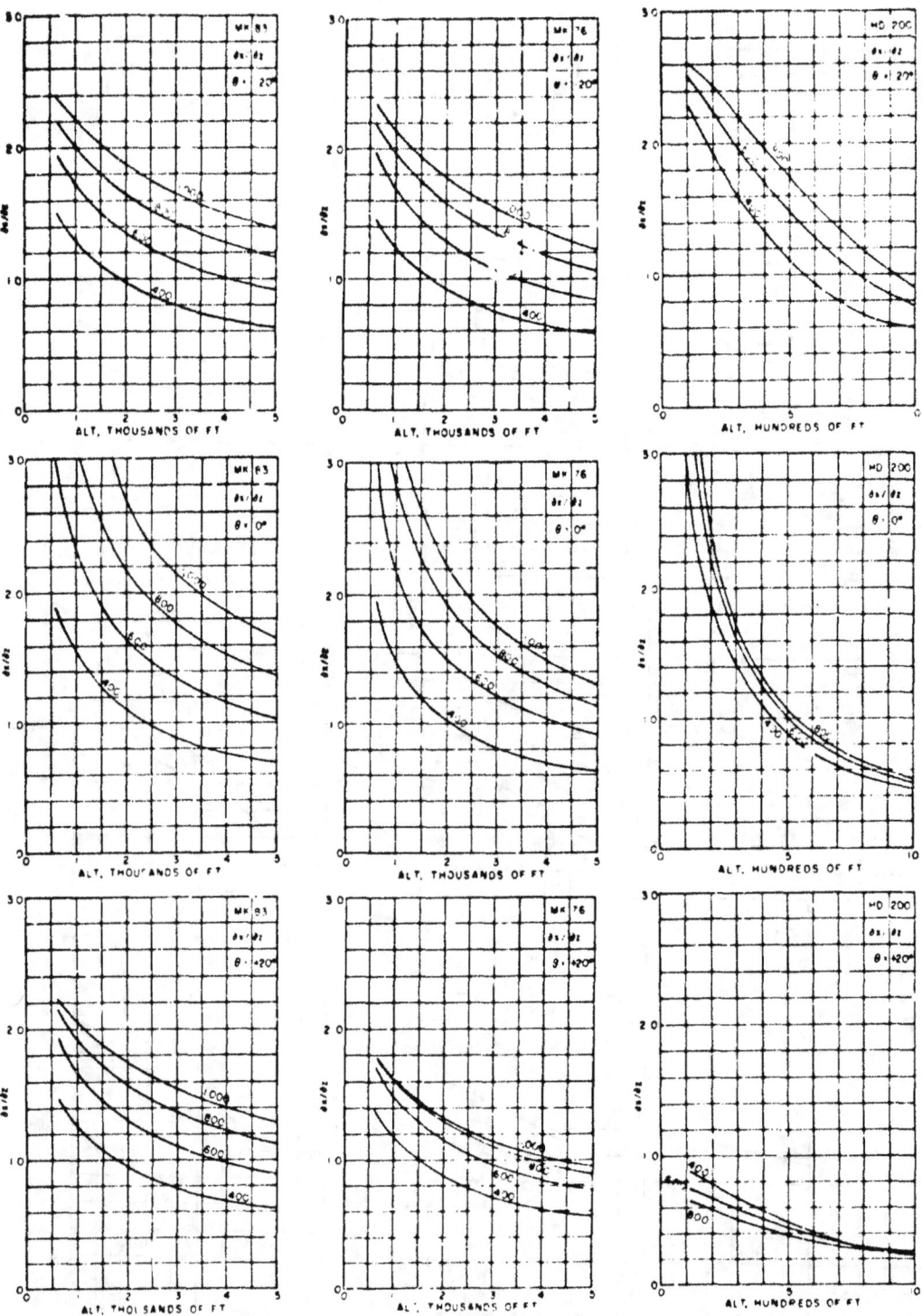

FIG. 25. Ground Range Sensitivity to Altitude Change Versus Altitude.

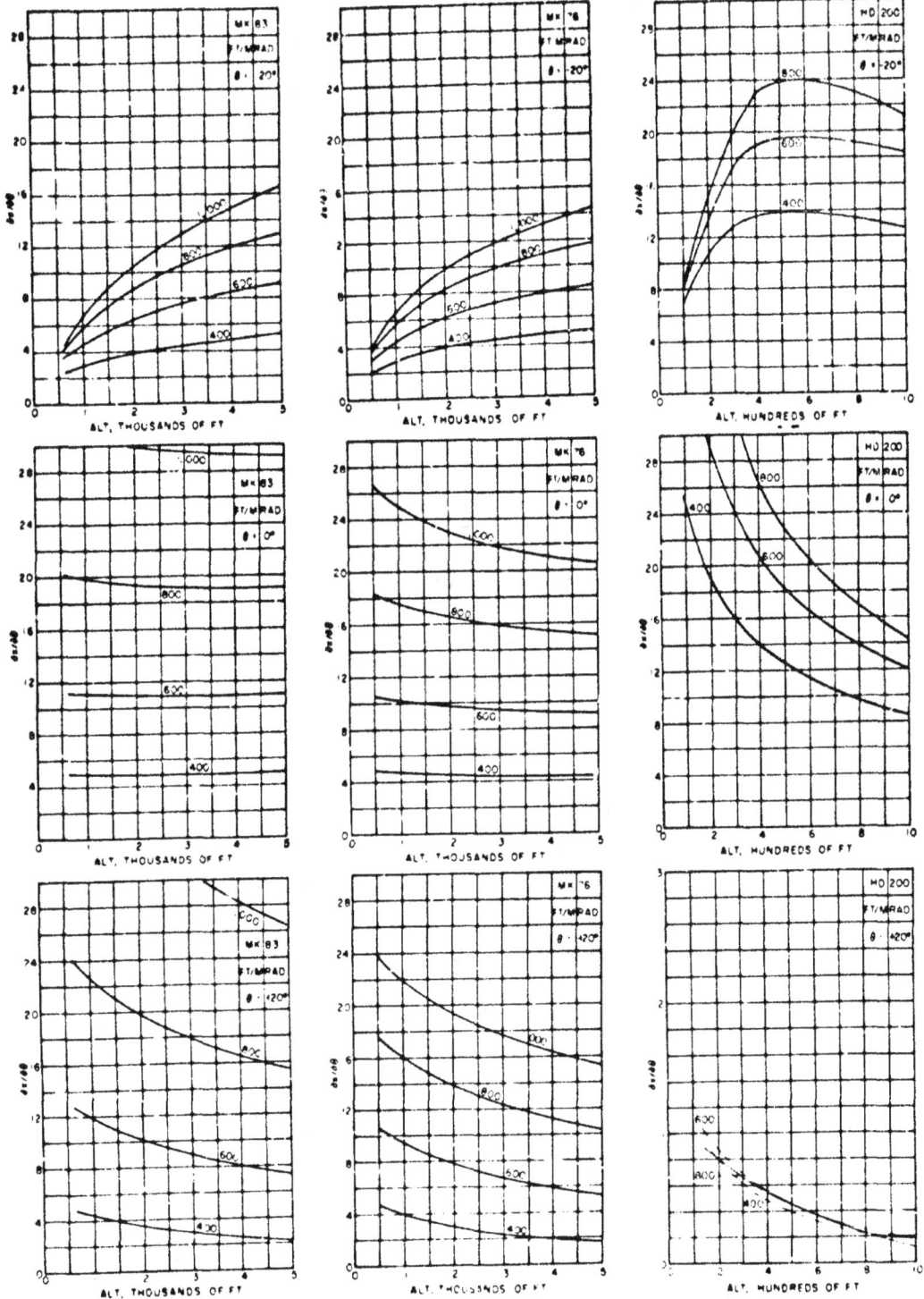

FIG. 26. Ground Range Sensitivity to Release Angle Change Versus Altitude.

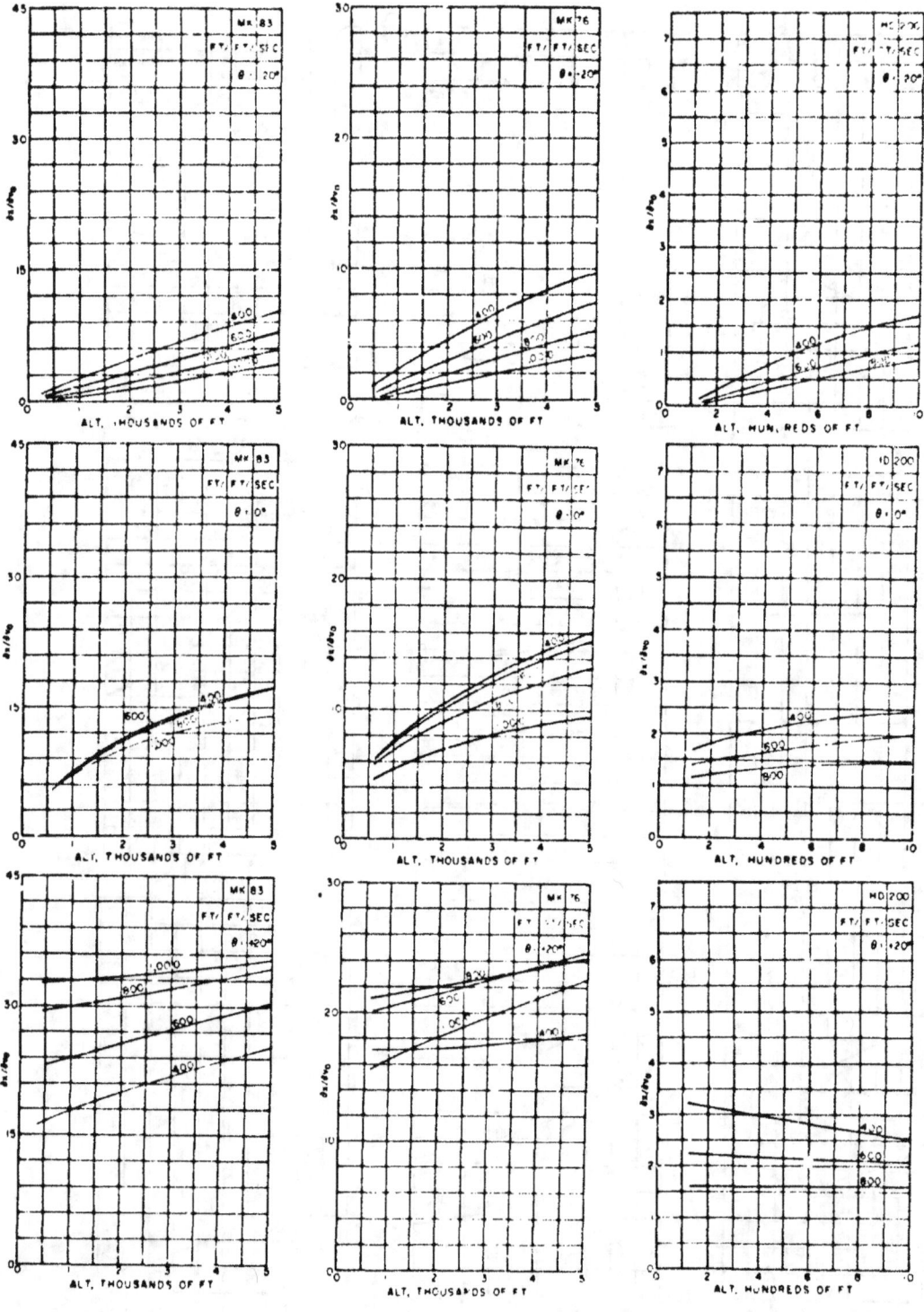

FIG. 27. Ground Range Sensitivity to Release Velocity Change Versus Altitude.

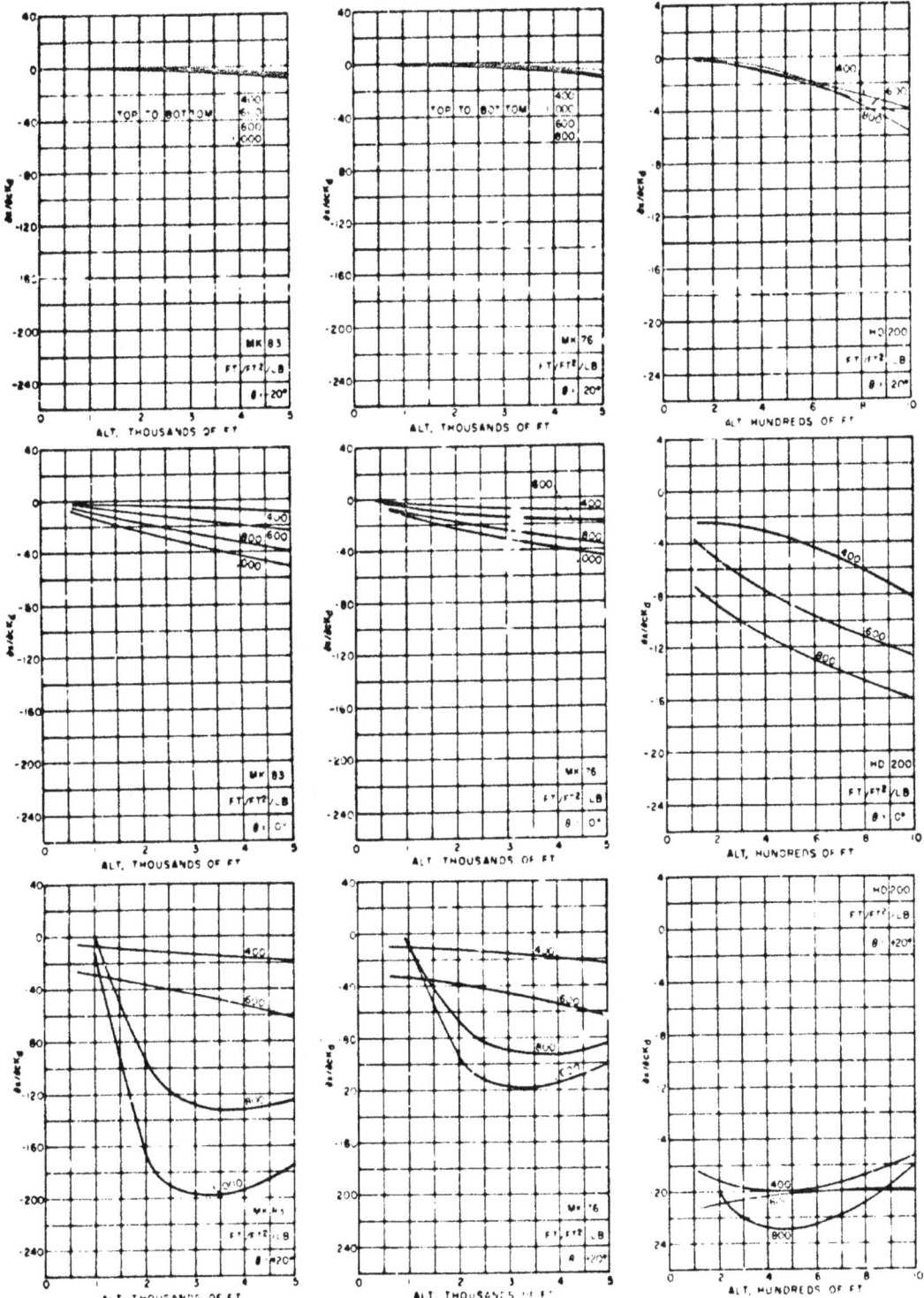

FIG. 28. Ground Range Sensitivity to Drag Function Change Versus Altitude.

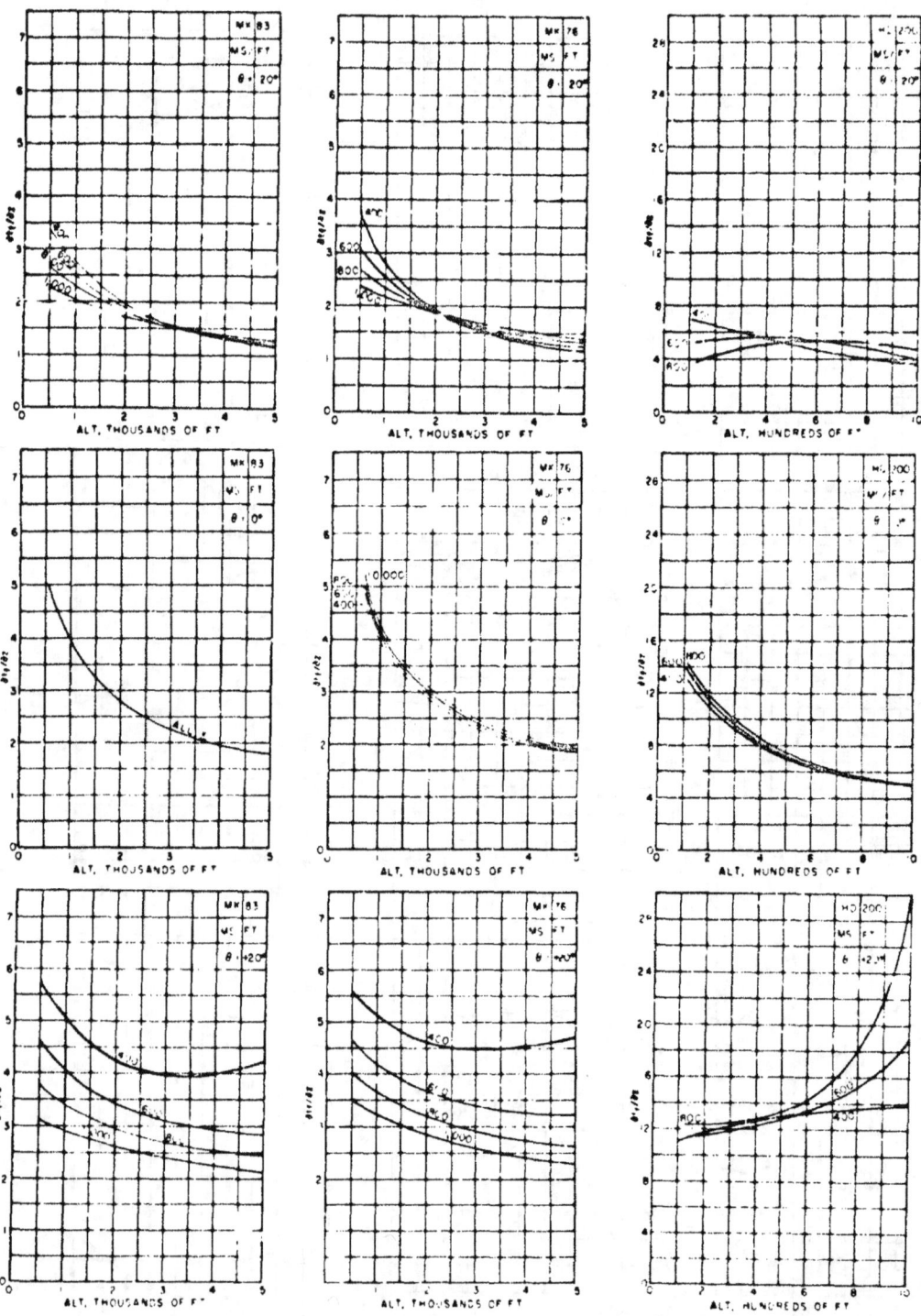

FIG. 29. Time of Flight Sensitivity to Altitude Change Versus Altitude.

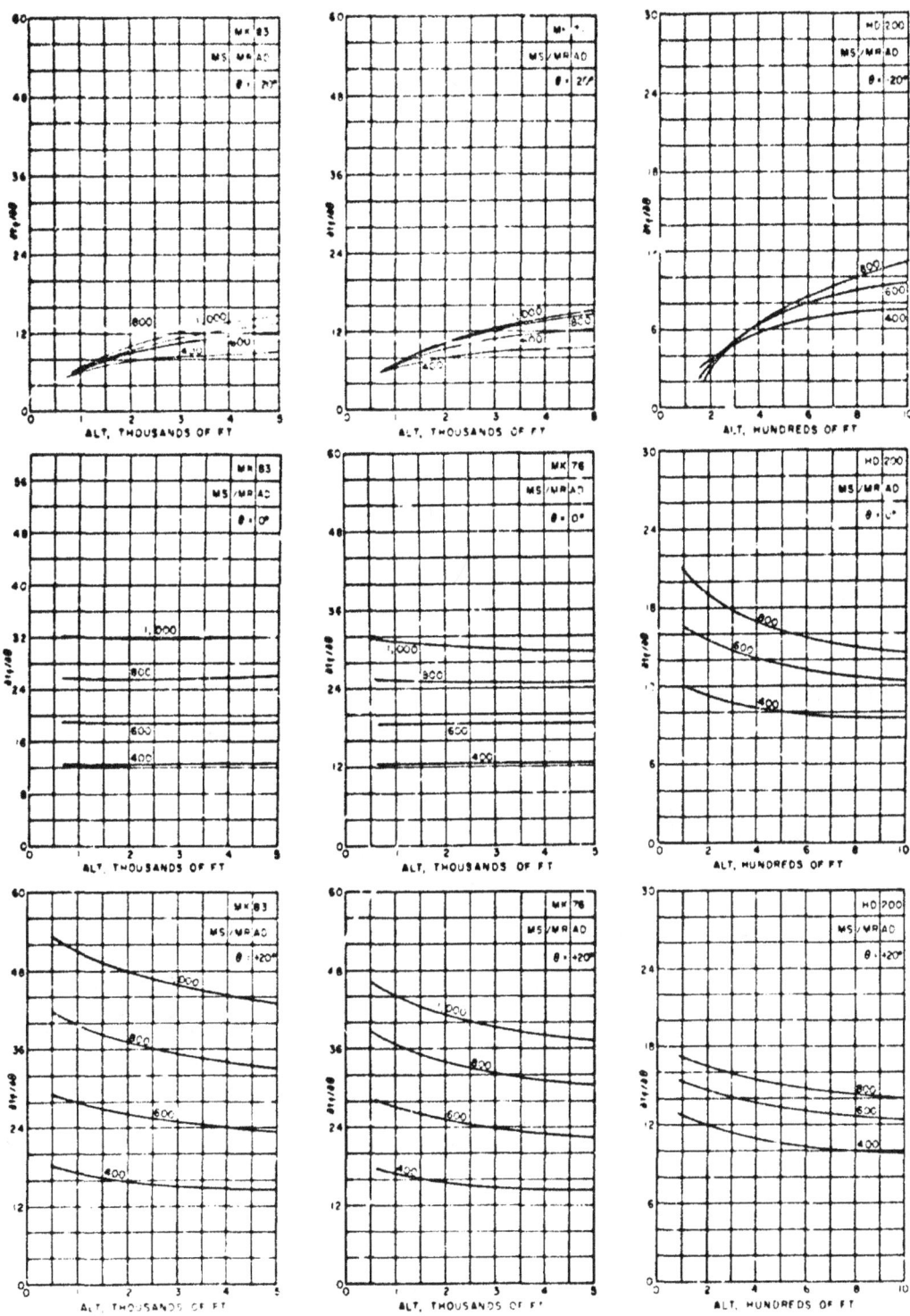

FIG. 30. Time of Flight Sensitivity to Release Angle Change Versus Altitude.

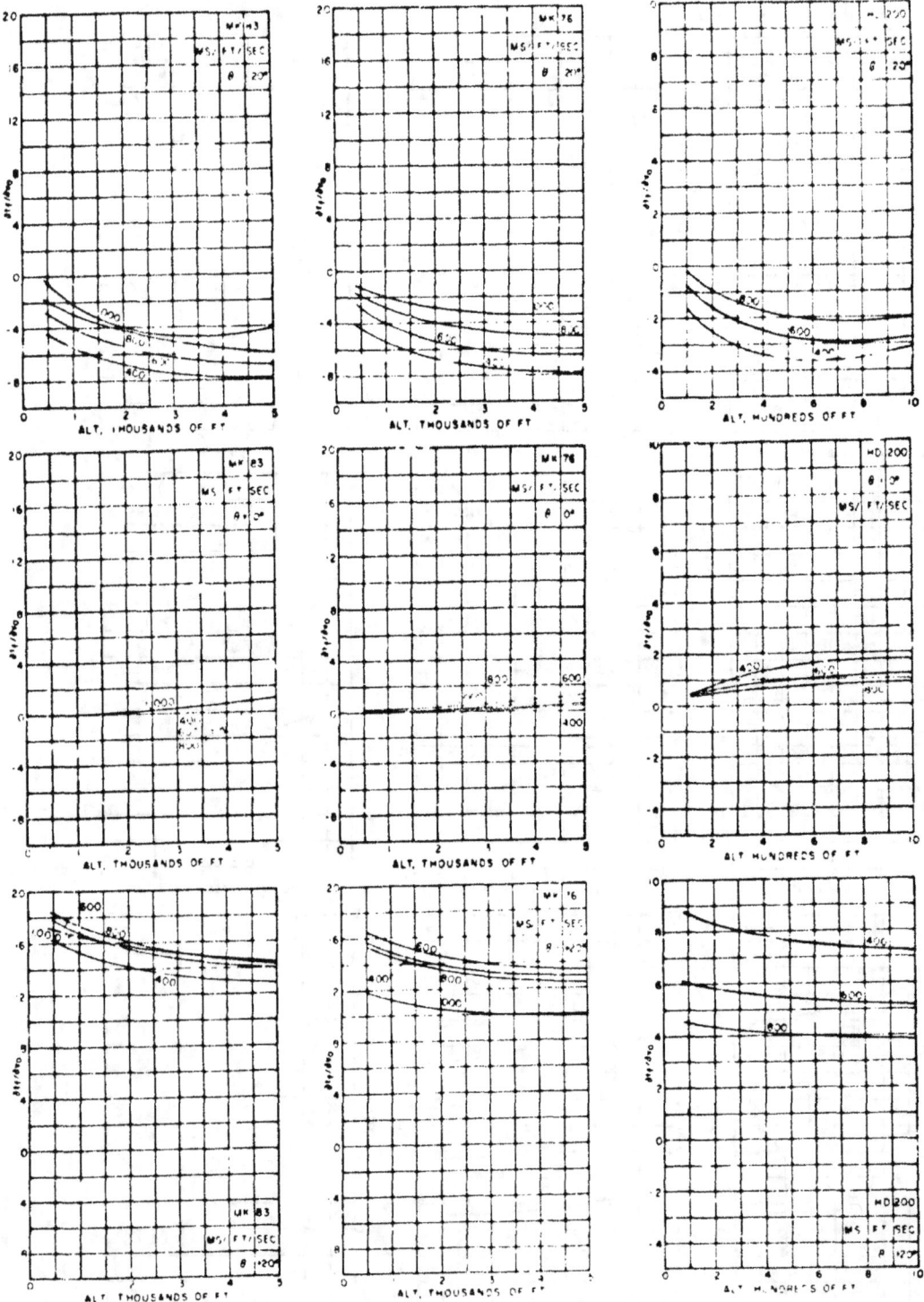

FIG. 31. Time of Flight Sensitivity to Release Velocity Change Versus Altitude.

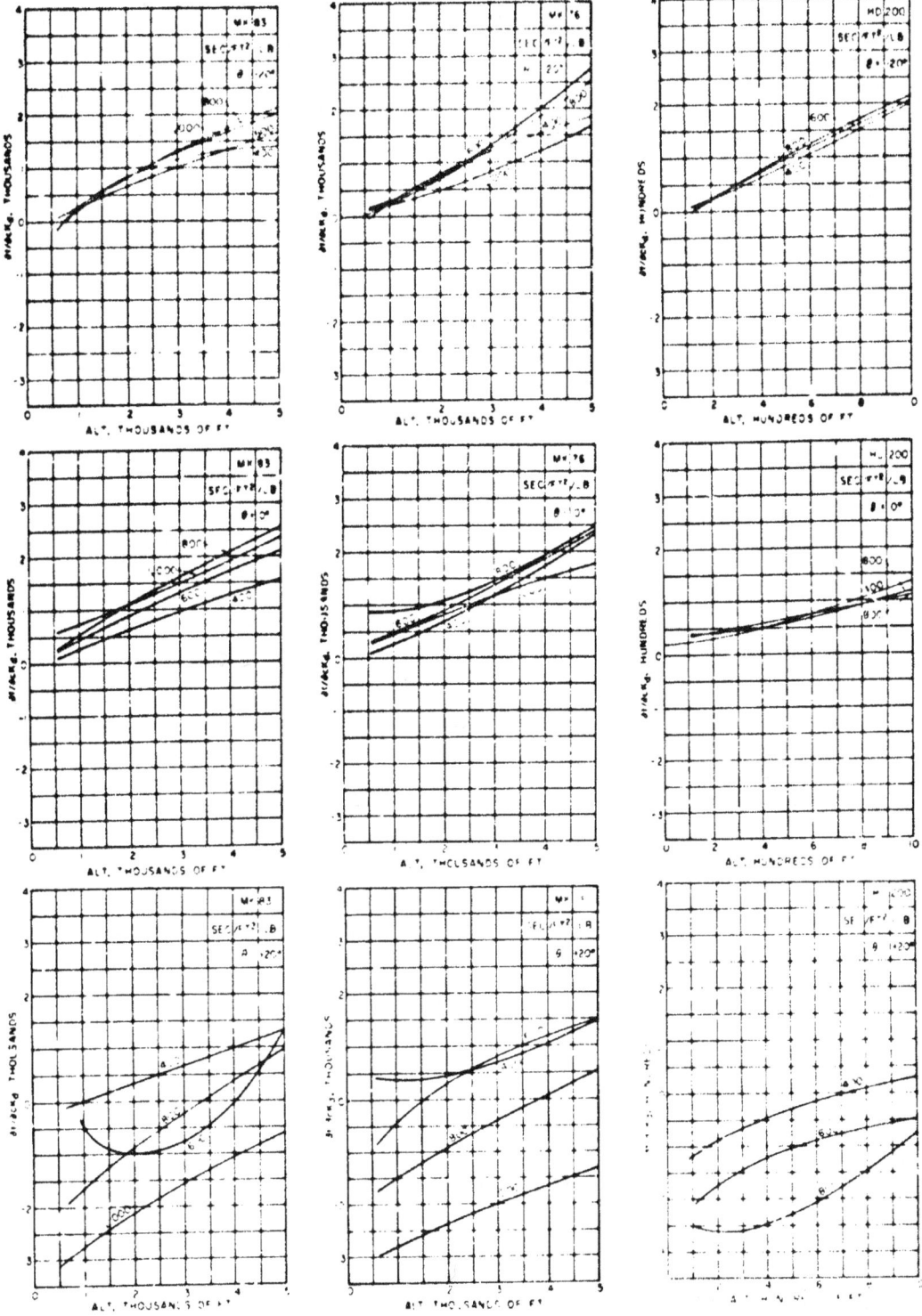

FIG. 32. Time of Flight Sensitivity to Drag Function Change Versus Altitude.

FIG. 33. Impact Angle Sensitivity to Altitude Change Versus Altitude.

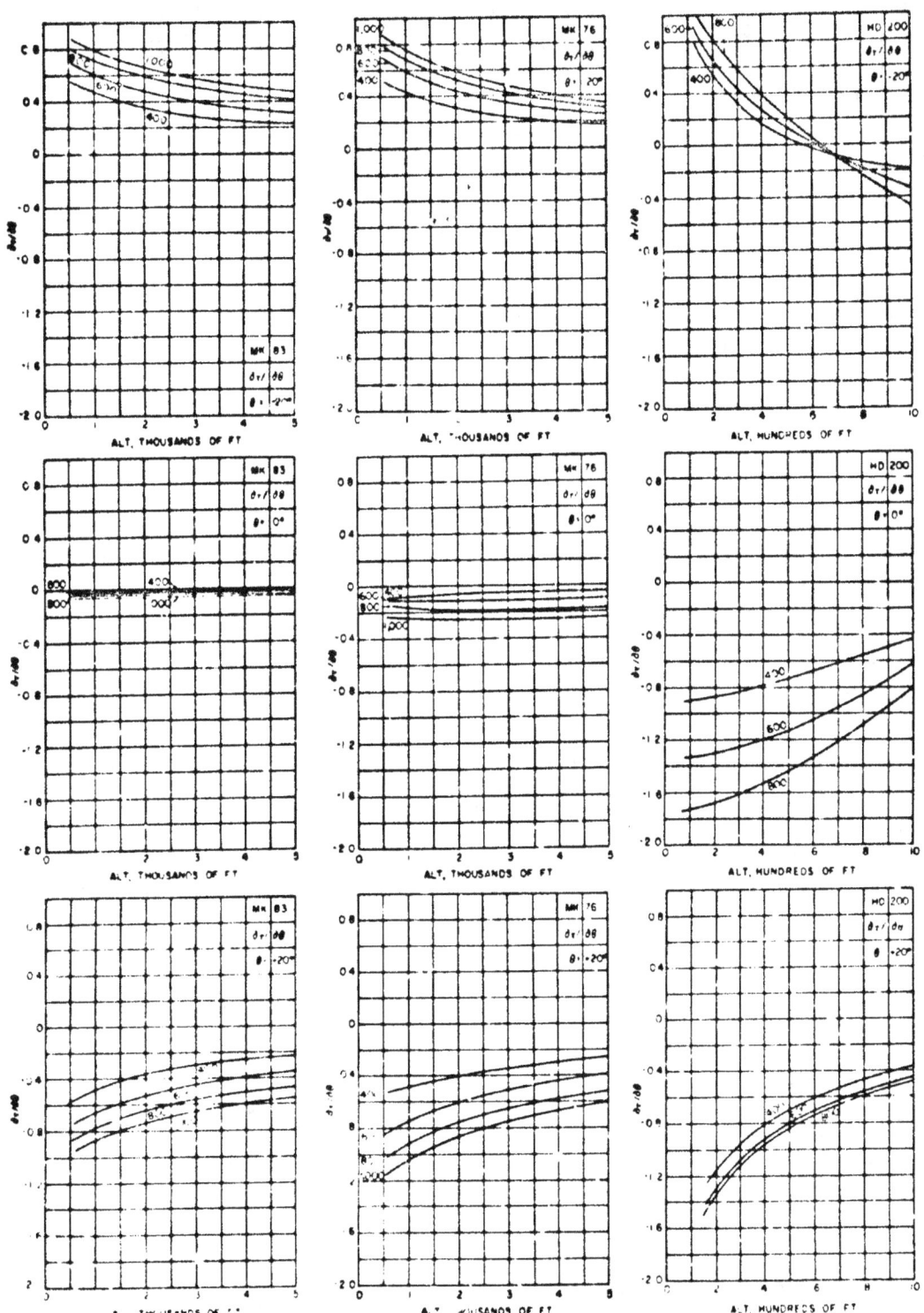

FIG. 34. Impact Angle Sensitivity to Release Angle Change Versus Altitude.

NOTS TP 3902

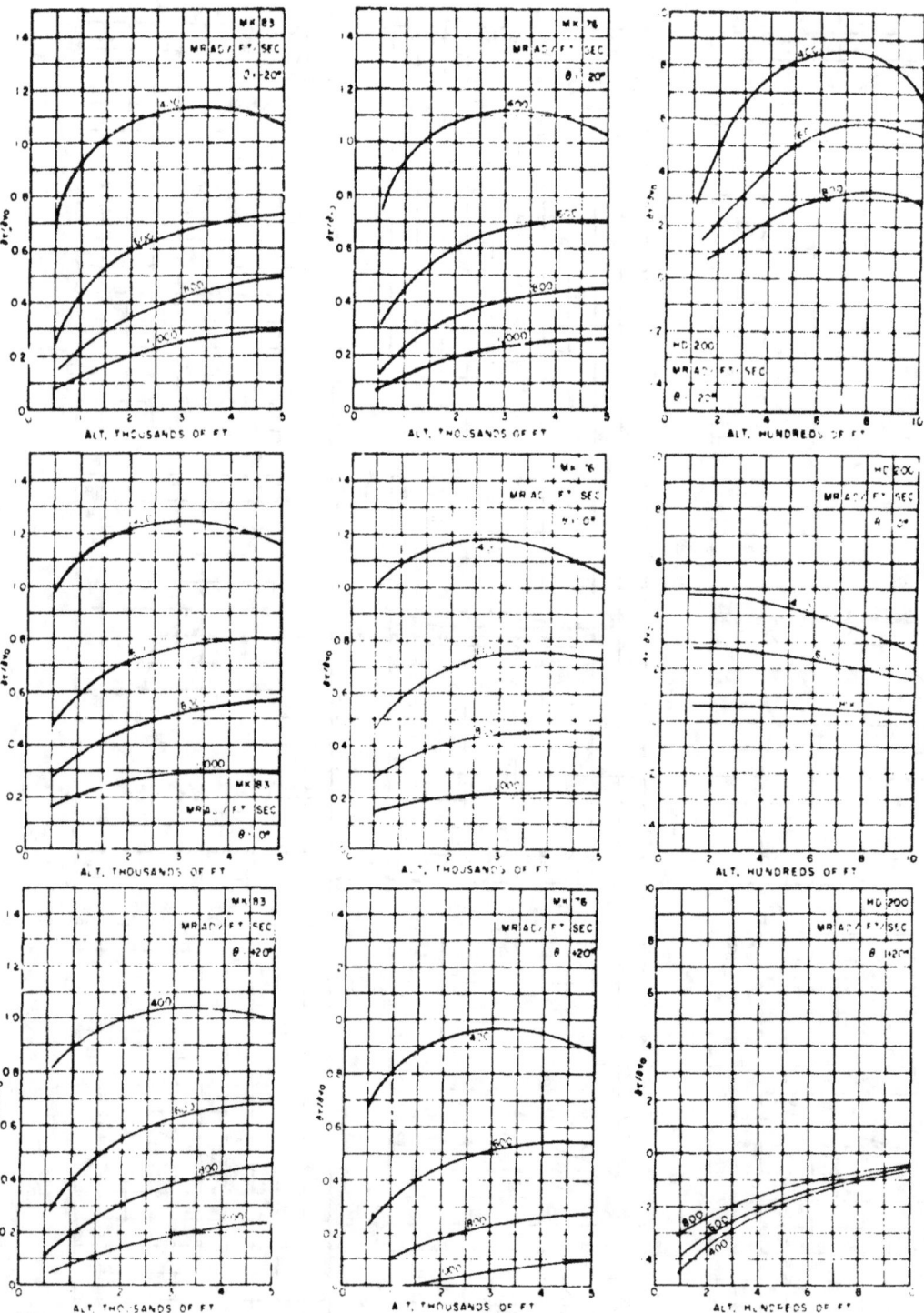

FIG. 35. Impact Angle Sensitivity to Release Velocity Change Versus Altitude.

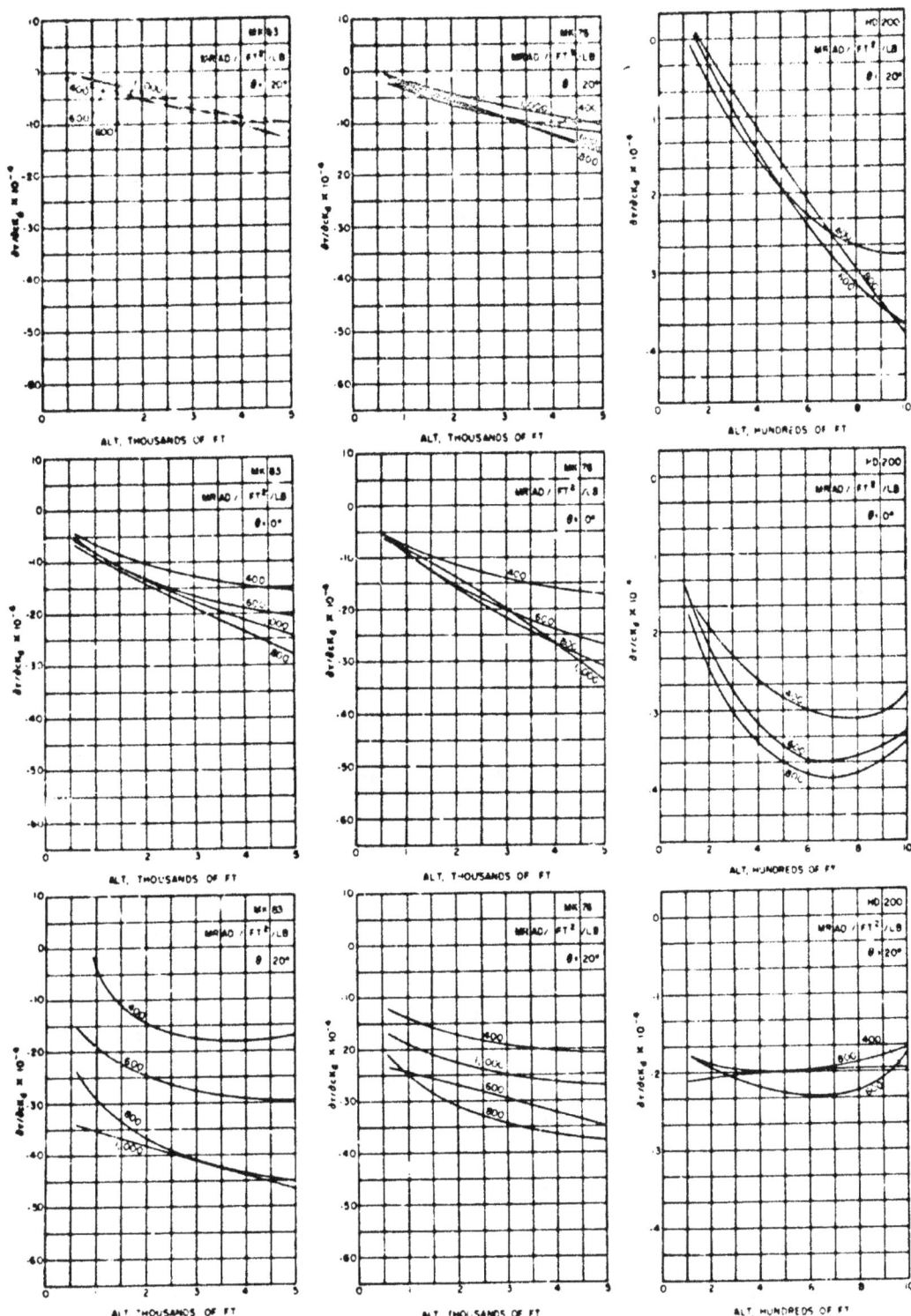

FIG. 36. Impact Angle Sensitivity to Drag Function Change Versus Altitude.

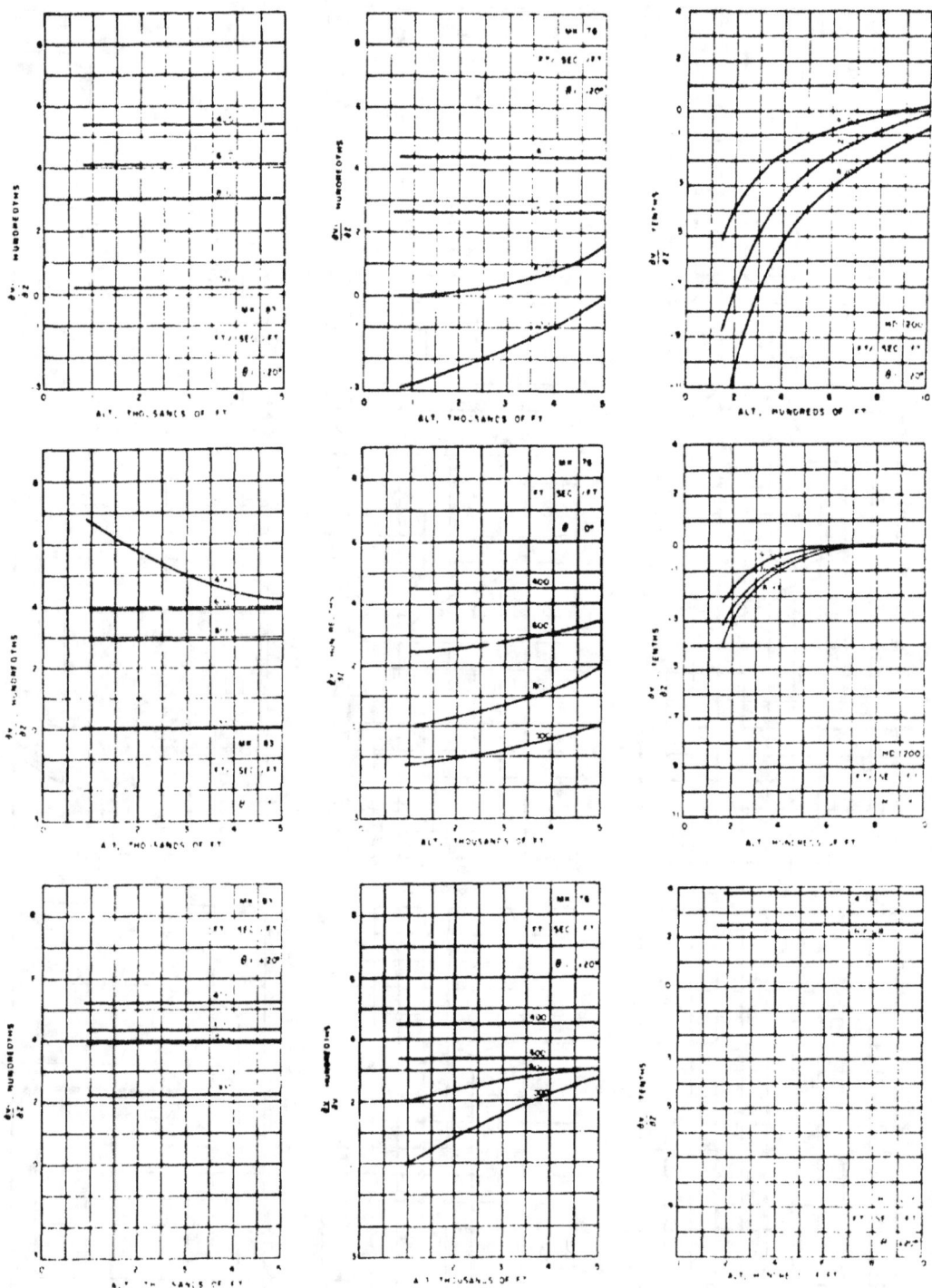

FIG. 37. Impact Velocity Sensitivity to Altitude Change Versus Altitude.

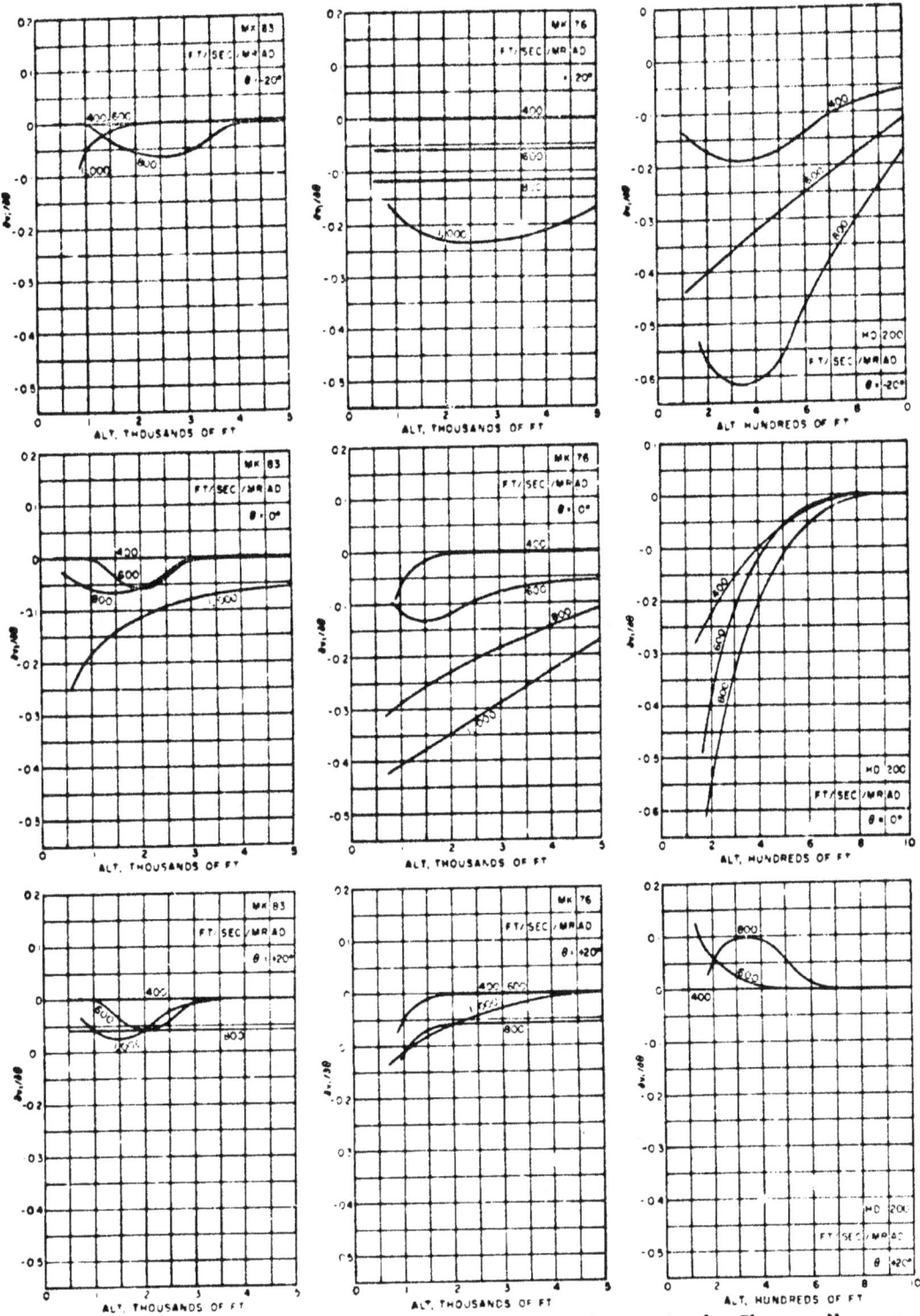

FIG. 38. Impact Velocity Sensitivity to Release Angle Change Versus Altitude.

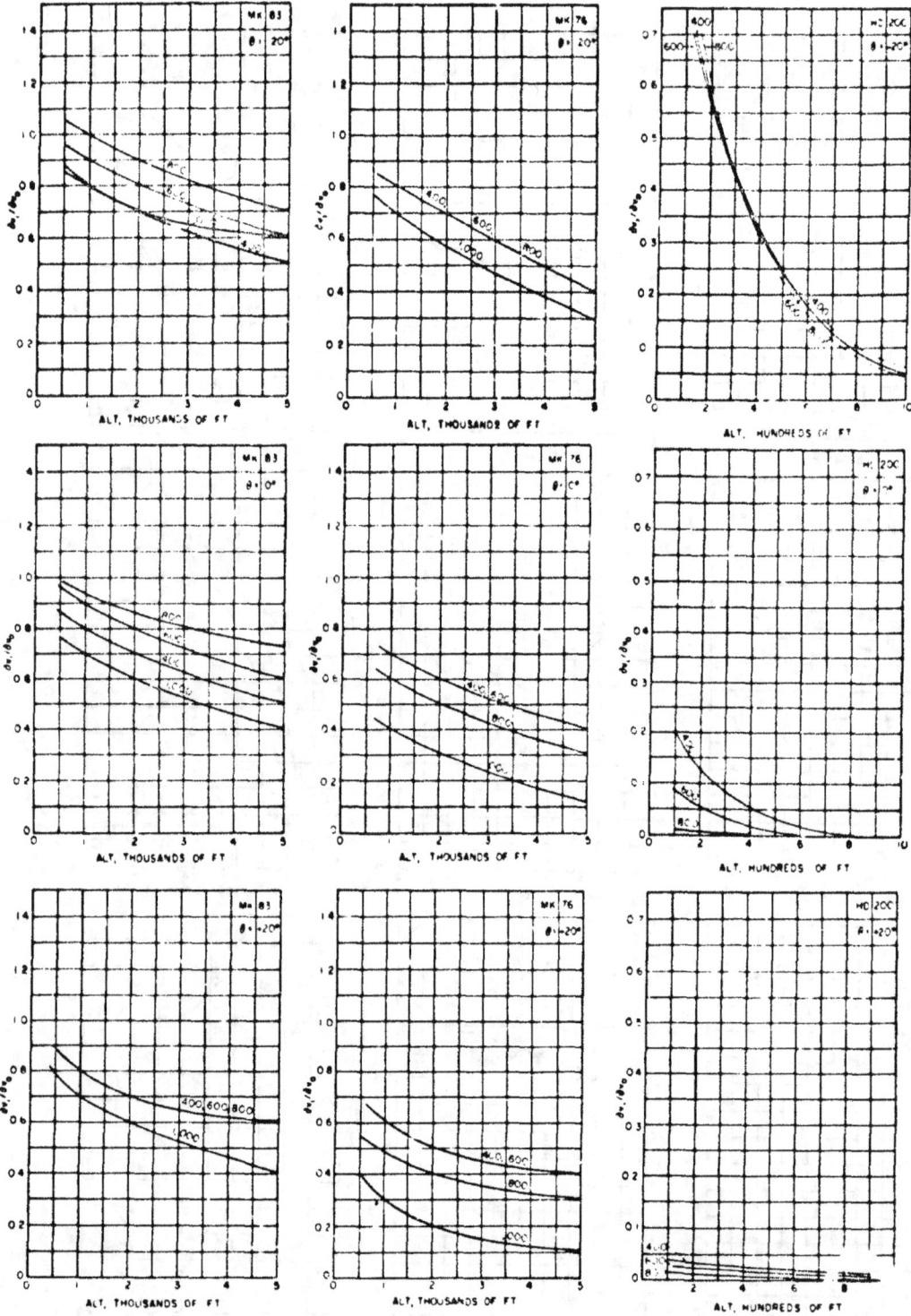

FIG. 39. Impact Velocity Sensitivity to Release Velocity Change Versus Altitude.

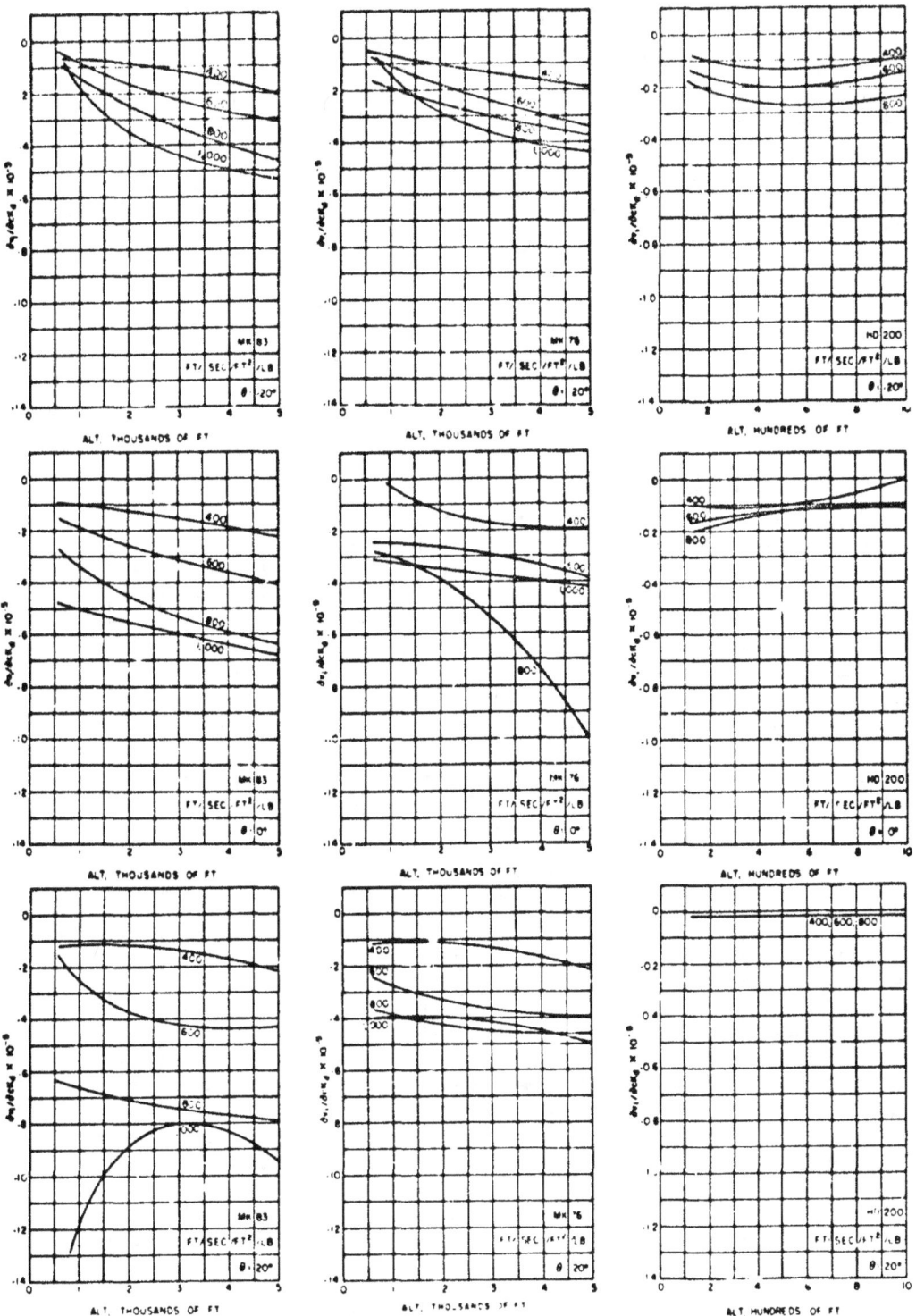

FIG. 40. Impact Velocity Sensitivity to Drag Function Change Versus Altitude.

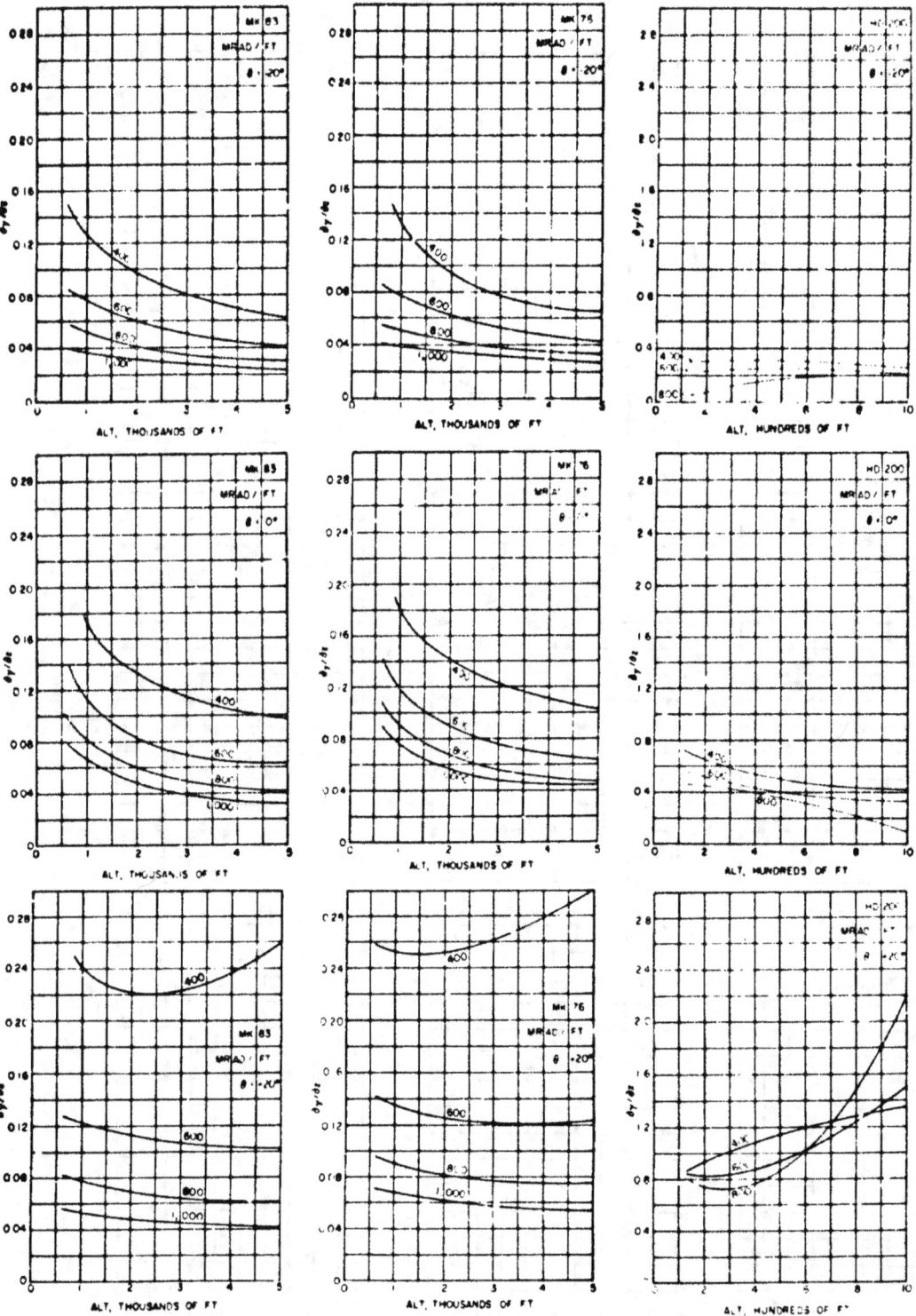

FIG. 41. Lead Angle Sensitivity to Altitude Change Versus Altitude.

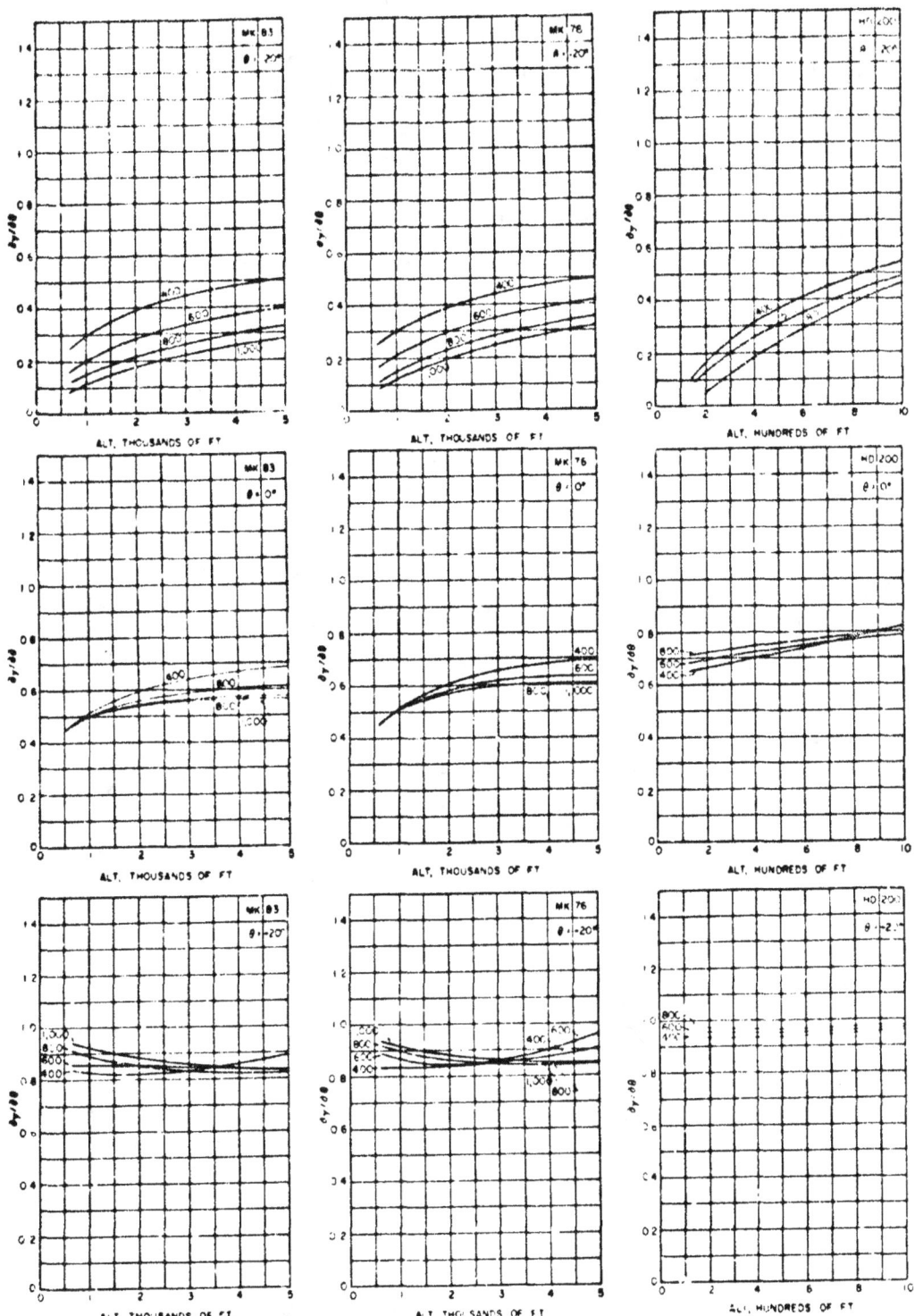

FIG. 42. Lead Angle Sensitivity to Release Angle Change Versus Altitude.

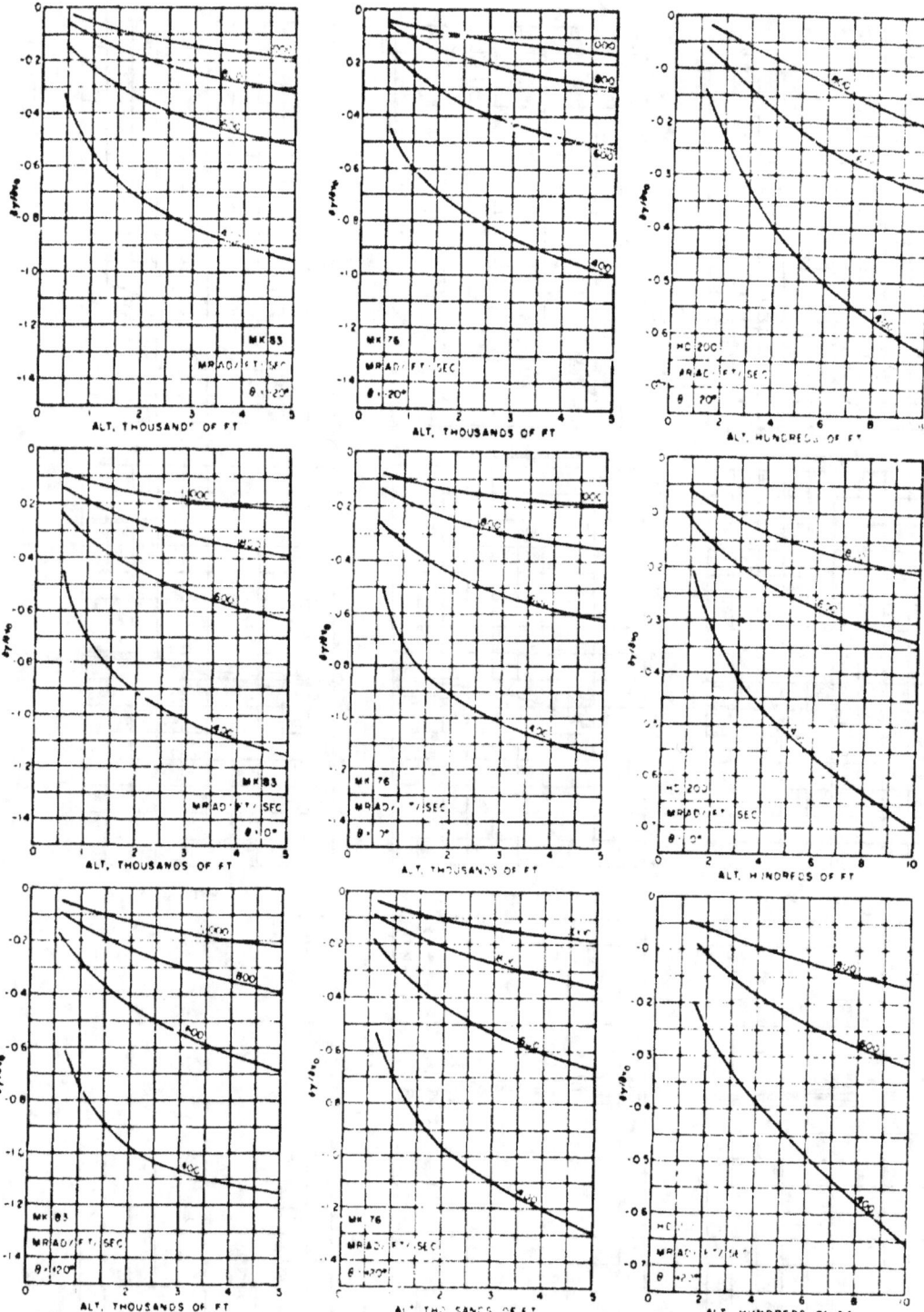

FIG. 43. Lead Angle Sensitivity to Release Velocity Change Versus Altitude.

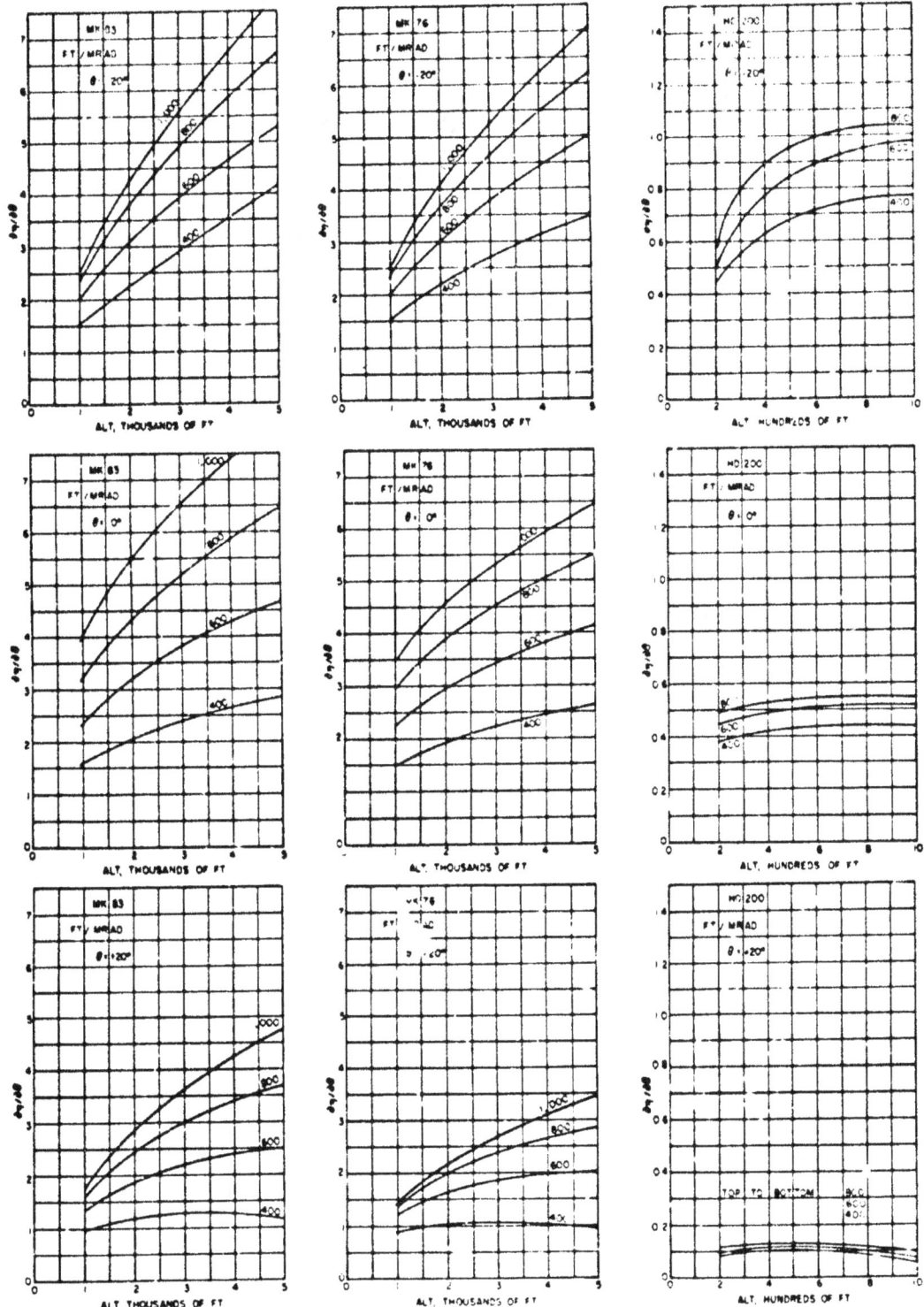

FIG. 44. Sensitivity of Distance Normal to Line-of-Sight to Release Angle Change Versus Altitude.

F. BOMB DATA

Table 10 gives some pertinent unclassified information on various bombs. The correct K_D function is given and the various K_D functions may be found tabulated in Section III.C. In most cases, several bombs can be made to use the same K_D curve by incorporating a correction factor in the value given for the reciprocal ballistic coefficient, c. Thus, c may include a form factor unequal to 1 (see Section II.B.3).

For reference to these bombs, a shortened designation system is used; i.e., the Mk 81 Mod 1 bomb with electric fuze is written Mk 81/1/E. For most bombs the mod number follows the first bar; after the second bar is given other pertinent conditions are given by the following abbreviations:

        T   without fuze
        E   electric fuze
        M   mechanical fuze
        L   with lugs
        N   without lugs
       WF water filled
       WS wet sand filled

TABLE 10. Bomb Data

Bomb	Mod	Diameter, in.	Weight, lb	Specifications	c ft^2/lb	K_D table to use
Mk 28	(EX)1	20.00	2040	–	.0008262	Mk76/4/T/N
Mk 43	0	18.00	2125	Large fin, nose Mod 1	.0010588	Mk43/0 nose Mod 1 (with large fin)
Mk 57	0, 1	14.75	500	–	.00244	"
Mk 76	0, 2	4.00	23.6	Without lugs	.004706	Mk76/0&2/N
Mk 76	2	4.00	23.6	With lugs	.006379	Mk76/0&2/N
Mk 76	4	4.00	24.23	With lug, no fuze	.004586	Mk76/4/T/L
Mk 76	4	4.00	24.23	Without lug or fuze	.004586	Mk76/4/T/N
Mk 81	1	9.00	270	Electric fuze	.002937	Mk83/2&3/E
Mk 81	1	9.00	270	Mechanical fuze	.003916	Mk83/2&3/E
Mk 82	0, 1	10.75	500	Electric fuze	.001814	Mk83/2&3/E
Mk 82	0, 1	10.75	500	Mechanical fuze	.002721	Mk83/2&3/E
Mk 83	2, 3	14.00	985	Electric fuze	.001382	Mk83/2&3/E
Mk 83	2, 3	14.00	985	Mechanical fuze	.001759	Mk83/2&3/E
Mk 84	1	18.00	1970	Electric fuze	.001142	Mk83/2&3/E
Mk 84	1	18.00	1970	Mechanical fuze	.001522	Mk83/2&3/E
Mk 86	0, 1	9.00	140	Water filled	.005936	Mk83/2&3/E
Mk 86	0, 1	9.00	200	Wet sand filled	.003994	Mk83/2&3/E
Mk 88	0	14.00	458	Water filled	.002972	Mk83/2&3/E
Mk 88	0	14.00	783	Wet sand filled	.001738	Mk83/2&3/E
Mk 89	0	4.00	56	Without lugs	.002877	Mk83/2&3/E
Mk 89	0	4.00	56	With lugs	.004250	Mk83/2&3/E
Mk 106	0	3.87	4.63	–	.1306	Std G1 drag function
Mk 106	2	3.87	4.65	–	.1511	"
HD-200		–	–	–	.1000	.1066 (constant) $M \leq 0.75$
AN-M57A1	–	10.80	284	M126 fin assembly	.002853	AN-M57A1 M126 fin
AN-M64A1	–	14.20	587	M128A1 fin assembly	.002385	AN-M64A1 M128A1 fin

IV. NOMOGRAPHS

The nomographs[5] given herewith allow rapid evaluation of many important parameters. While they cannot give the accuracy of a computed value, they can give an answer that is well within allowable limits of error (usually less than one or two percent) for design and analysis of bombs and types of fire control systems.

Many of the nomographs are versatile; on some of the graphs either of two variables may be given and the remaining one calculated. They are most accurate near the center of their ranges; on some of the graphs the error will increase as velocities tend toward the speed of sound. The emphasis is on subsonic application throughout.

On all nomographs the examples given are drawn as dotted lines on the graphs. In some cases, a small diagram is given and on the most involved nomographs detailed instructions are included.

The nomographs are divided into four sections, which are described below. The first section contains graphs applicable to all bombs; the second contains vacuum solutions of the trajectory which might be useful in allowing for delay times; the third section deals with graphs applicable only to standard drag bombs; and the last section is restricted to retarded bombs.

It should be noted that many of the standard drag bombs require knowledge of the cK_D product. In such cases, a small copy of nomograph 7 is included; Mach numbers may be obtained from nomographs 7 <u>or</u> 16.

---

[5] Upon request, NOTS, China Lake, will provide specific nomographs on plastic.

NOTS TP 3902

DESCRIPTION OF NOMOGRAPHS

    A. GENERAL USAGE (p. 88)

        1. Altitude, or ground range (three nomographs in one)
        2. Altitude, or ground range
        3. Slant range

    B. VACUUM SOLUTIONS (p. 94)

        4. Ground range
        5. Impact angle
        6. Altitude

    C. STANDARD DRAG BOMBS (p. 100)

        7. Mach number, K_D, cK_D product, altitude corrections
        8. Impact angle
        9. Time of flight, or ground range
        10. Altitude
        11. Altitude or ground range, level release
        12. Ground range, loft bombing from $Z = 0$ to $Z = 0$
        13. Ballistic lead angle using ground range data
        14. Ballistic lead angle using altitude data
        15. Change in ground range due to small change in angle of release about level release

    D. RETARDED BOMBS (p. 118)

        16. Mach number, K_D, cK_D product, altitude corrections
        17. Impact angle
        18. Time of flight, or ground range
        19. Altitude
        20. Ground lag, ground range, or altitude, level release
        21. Ballistic lead angle using ground range data
        22. Change in ground range due to small change in angle of release about level release.
        23. Time of flight or altitude, level release

A. GENERAL USAGE

Nomograph 1. Altitude, or Ground Range

This nomograph solves the relations

$$Z = R \sin \phi,$$
$$Z = X \tan \phi,$$
$$X = R \cos \phi.$$

Given any two of the variables, R, X, Z, or ϕ, the remaining two can be determined.

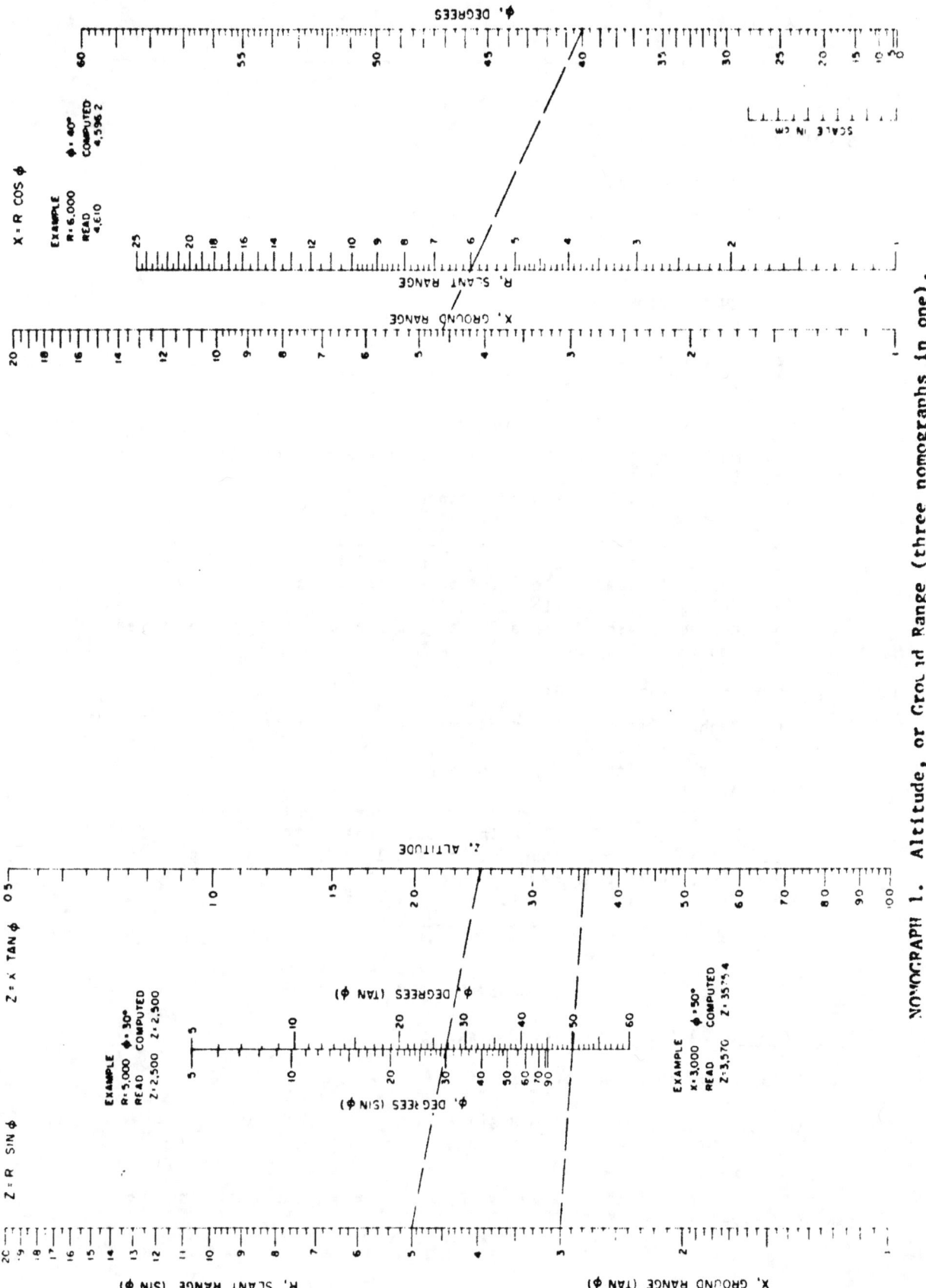

NOMOGRAPH 1. Altitude, or Ground Range (three nomographs in one).

NOTS TP 3902

Nomograph 2. Altitude or Ground Range (General Usage)

Given Z, X may be solved for using this graph. The process may be reversed to solve for Z although an iteration process may be necessary. This nomograph solves eq. 51a. Either nomograph 7 or 16 may be used to determine the cK_D value, depending on the type of bomb being studied.

Use of nomograph:

1. Locate release angle on upper left scale, label as A.
2. Locate velocity on lower left scale, label as B.
3. Construct line AB, label as C the point where AB crosses upper oblique index.
4. Locate altitude on far left scale, label as D.
5. Construct line CD, label as E point where CD cuts center vertical index.
   Steps 1 through 5 may be carried out on smaller scales at lower right center; this will give the same point E.
6. Locate release angle on scales in lower right, label as point F. Note that there are different scales depending on sign of θ.
7. Locate velocity on scales in lower right corner. Note that different U-scales are used for different signs of θ.
8. Construct line FG, label as H intersection of FG and left vertical index line. Construct line EH.
9. Locate θ on upper right scale, label as J.
10. Locate Z on upper right scale, label as K.
11. Construct line JK, label as L intersection of JK with index line. Locate and label as M the correct cK_D value.
12. Construct line LM, label as N the intersection of LM and upper left vertical index line.
13. From N, trace along curve to the left where curve intersects line EH; note value of ground range.

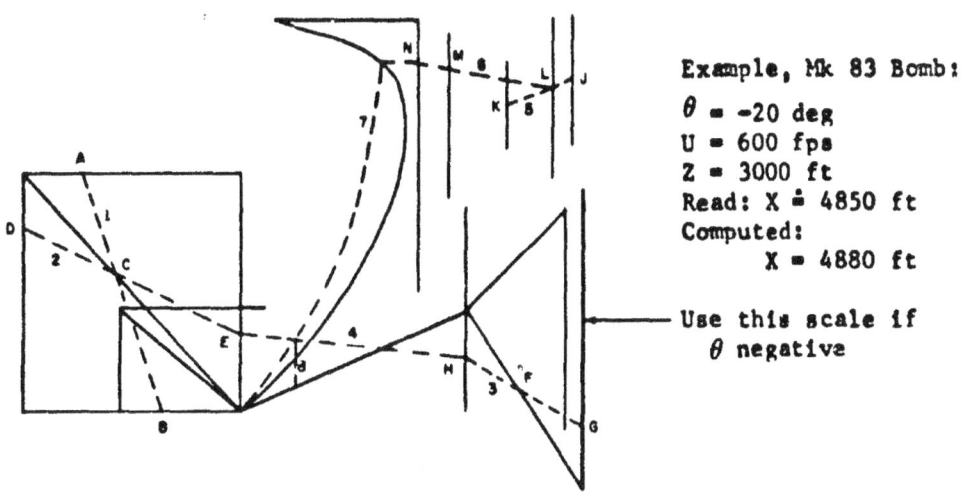

Example, Mk 83 Bomb:
$\theta = -20$ deg
$U = 600$ fps
$Z = 3000$ ft
Read: $X \doteq 4850$ ft
Computed:
$X = 4880$ ft

Use this scale if θ negative

$$z = -x \tan \theta + \frac{g x^2 \psi}{2 u^2 \cos^2 \theta}$$

$$\psi = \psi (kx \sec \theta)$$

$$k = \frac{2}{3} \rho c K_D$$

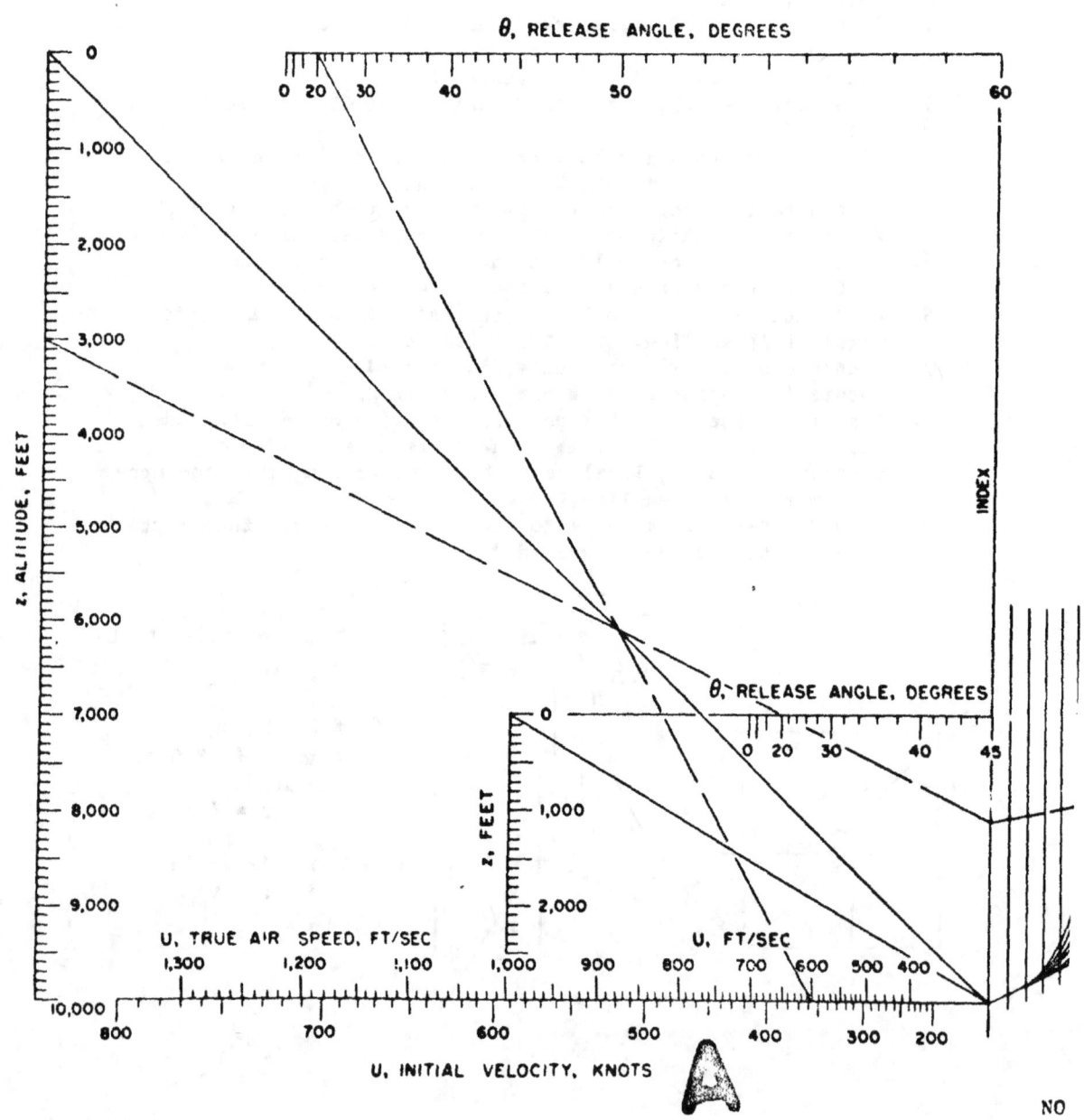

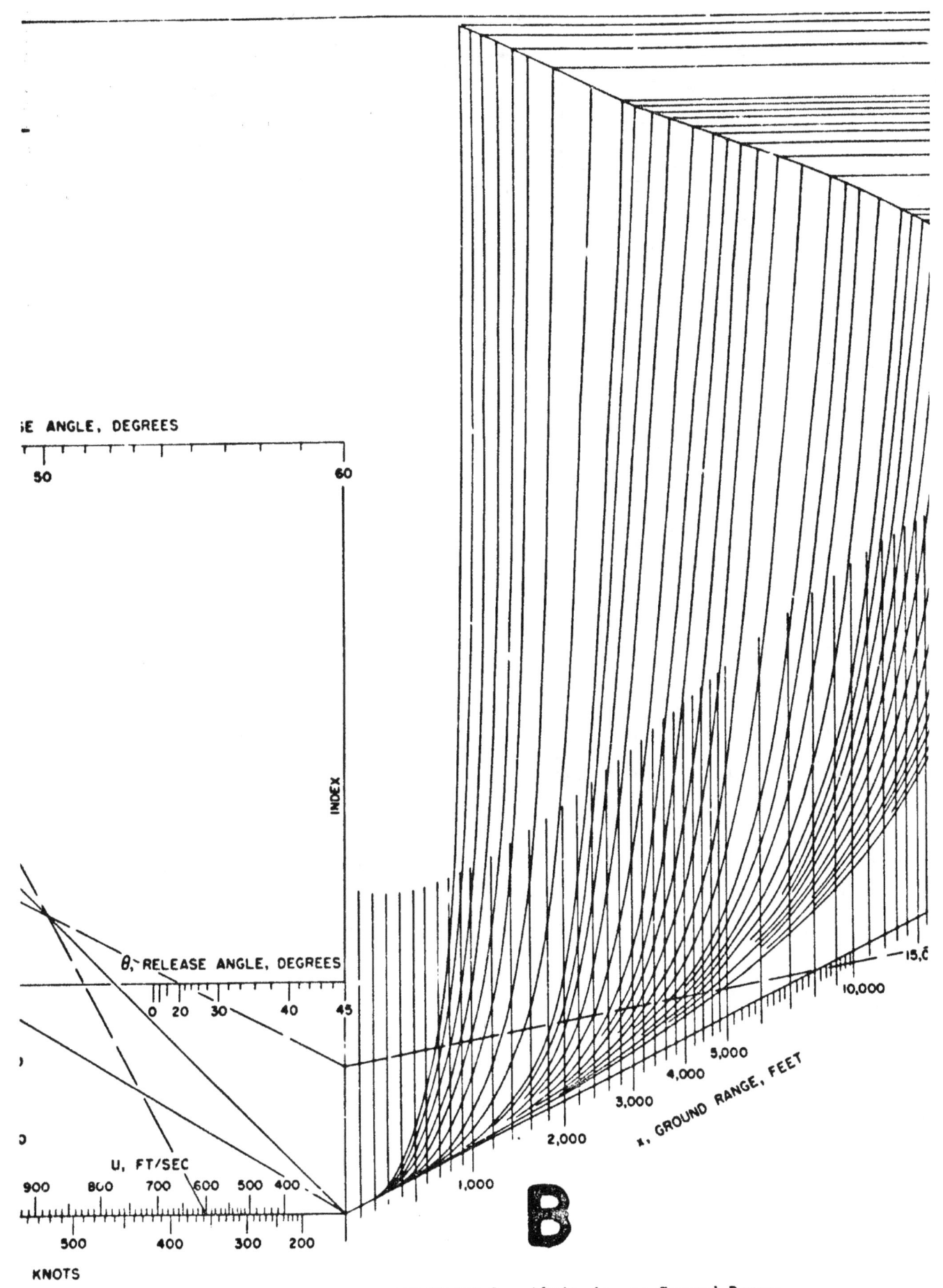

NOMOGRAPH 2. Altitude, or Ground Range.

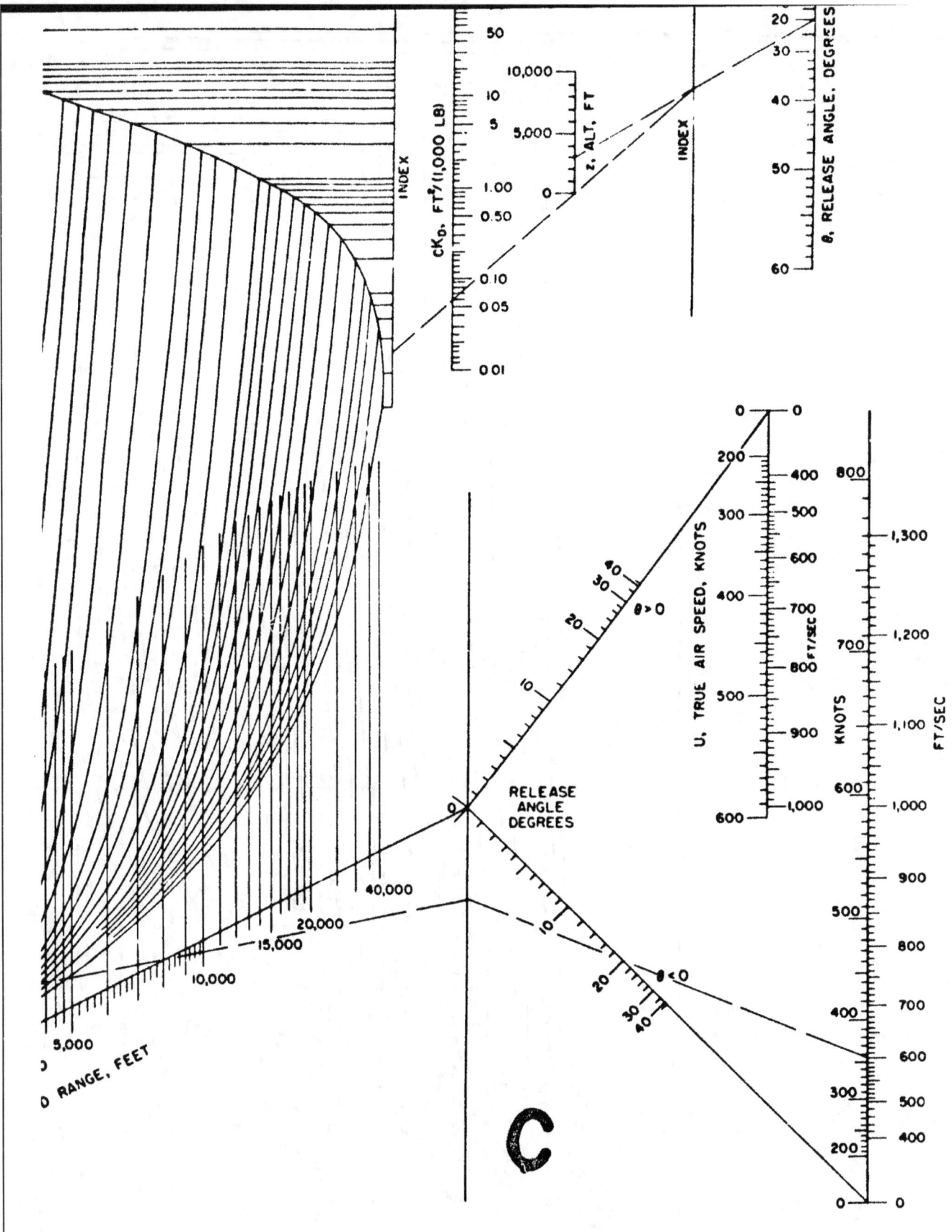

Nomograph 3. Slant Range (General Usage)

This nomograph solves the relation

$$R^2 = X^2 + Z^2$$

for any of the variables R, X, or Z, given the other two.

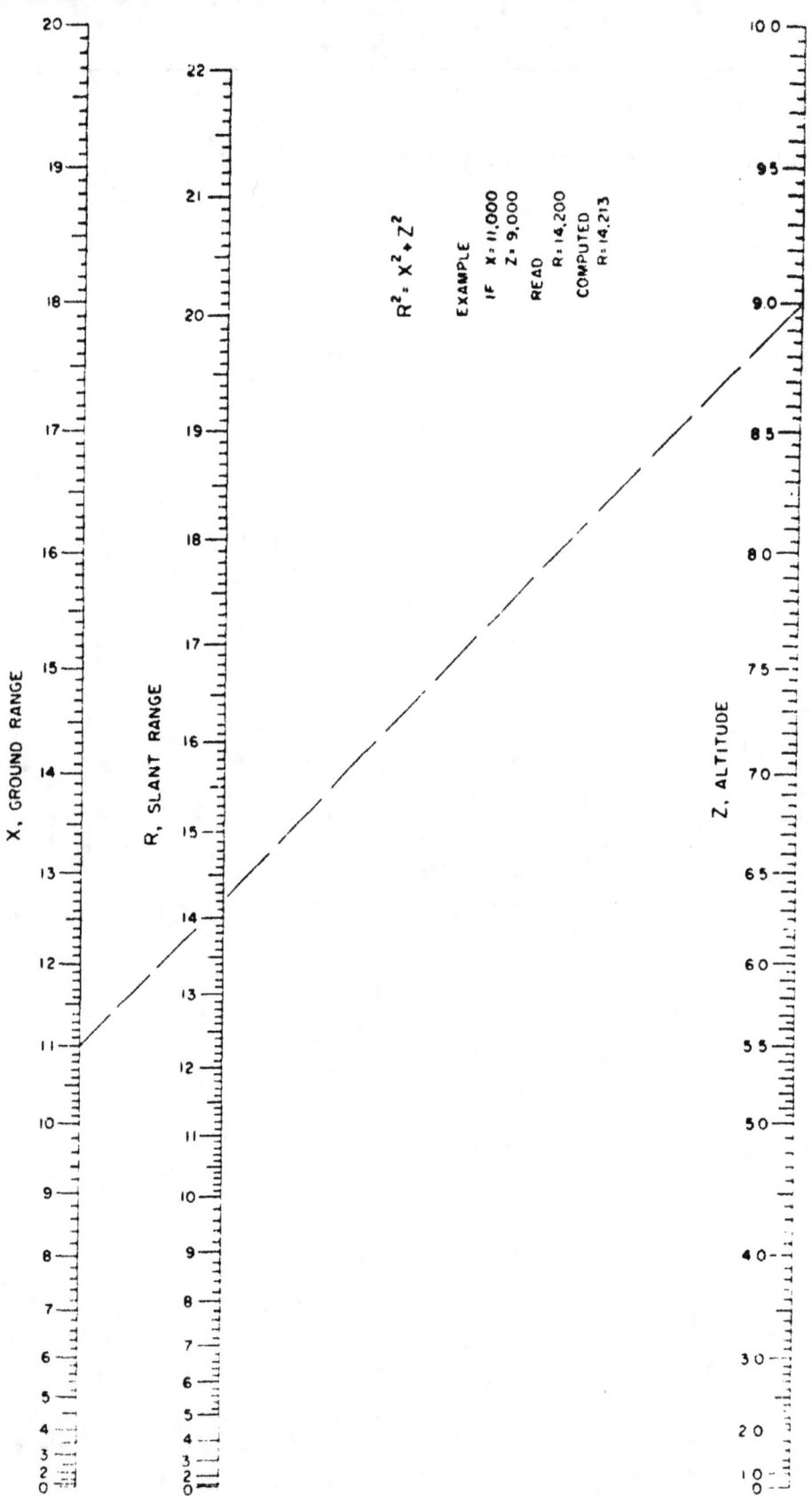

NOMOGRAPH 3. Slant Range.

B. VACUUM SOLUTIONS

Nomograph 4. Ground Range

This nomograph solves the relation

$$X = Ut \cos \theta .$$

Given any three of the variables X, U, t, or θ, the fourth may be found merely by interchanging, as required, the steps indicated on the nomograph (use diagram).

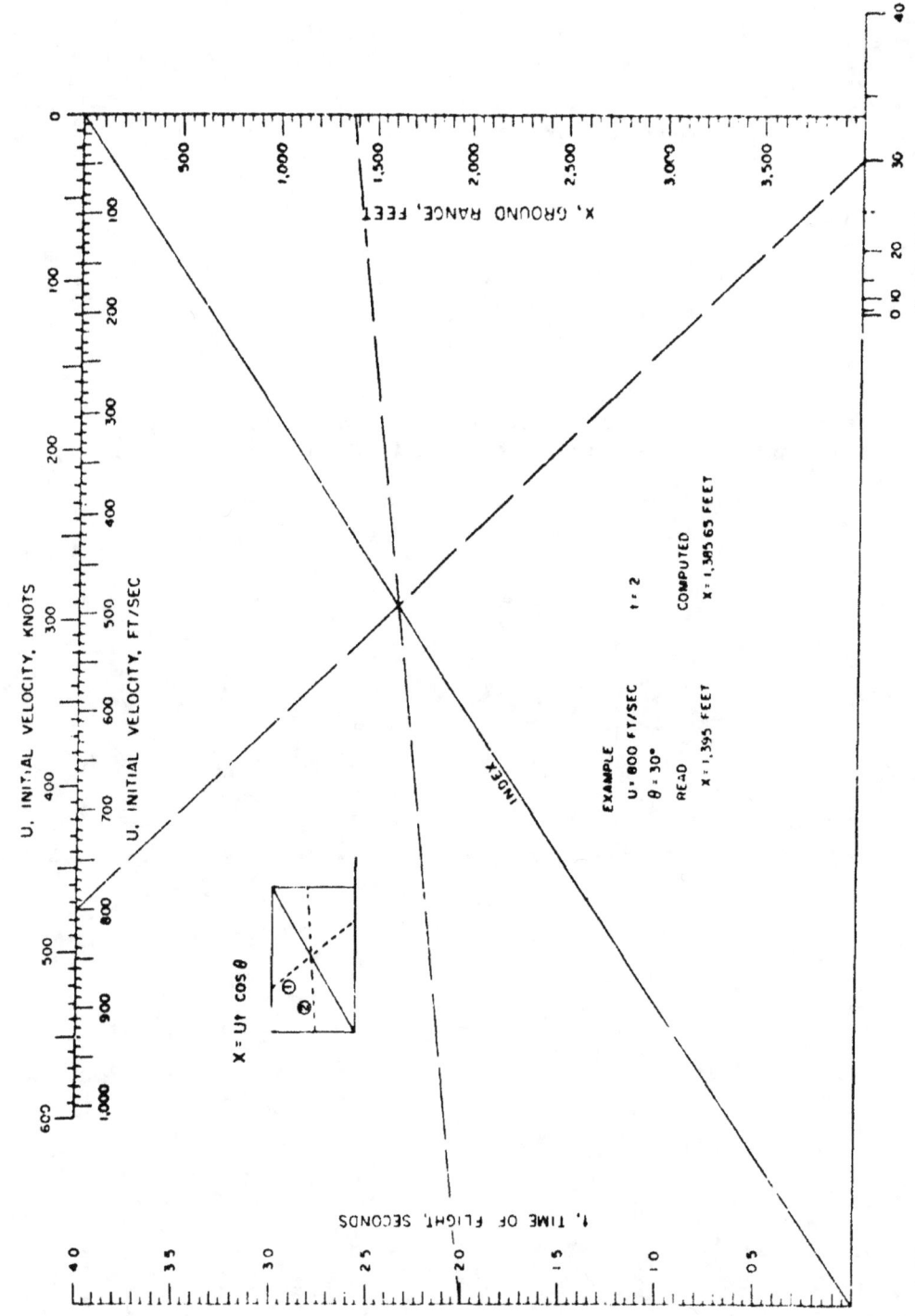

NOMOGRAPH 4. Ground Range.

Nomograph 5. Impact Angle (Vacuum Solutions)

Using the equations for the trajectory in a vacuum, this nomograph allows rapid determination of impact angle. The nomograph is based on eq. 26.

For standard drag bombs, this might serve as a very rough approximation of the actual conditions; other nomographs of this section might also be applicable to solution of problems involving delay times of retarded bombs.

Use of Nomograph 5: Given θ, U, and t, enter the nomograph as indicated by the nomograph diagram below to find the impact angle.

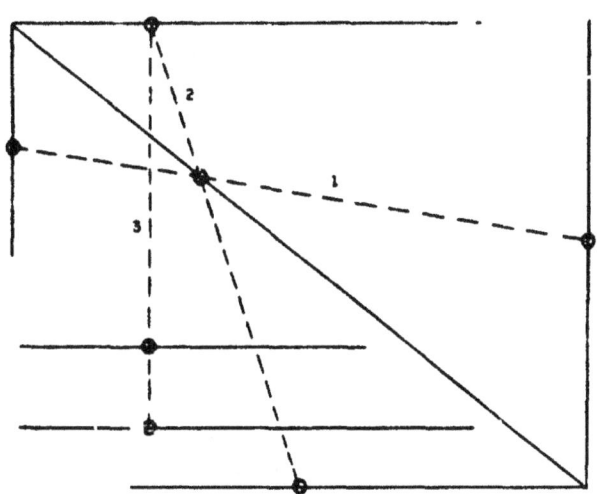

Example:

θ = -20 deg
U = 500 fps
t = 2 sec (time of flight)

Read: $\tau \doteq$ 26.6 deg
Computed: τ = 26.61 deg

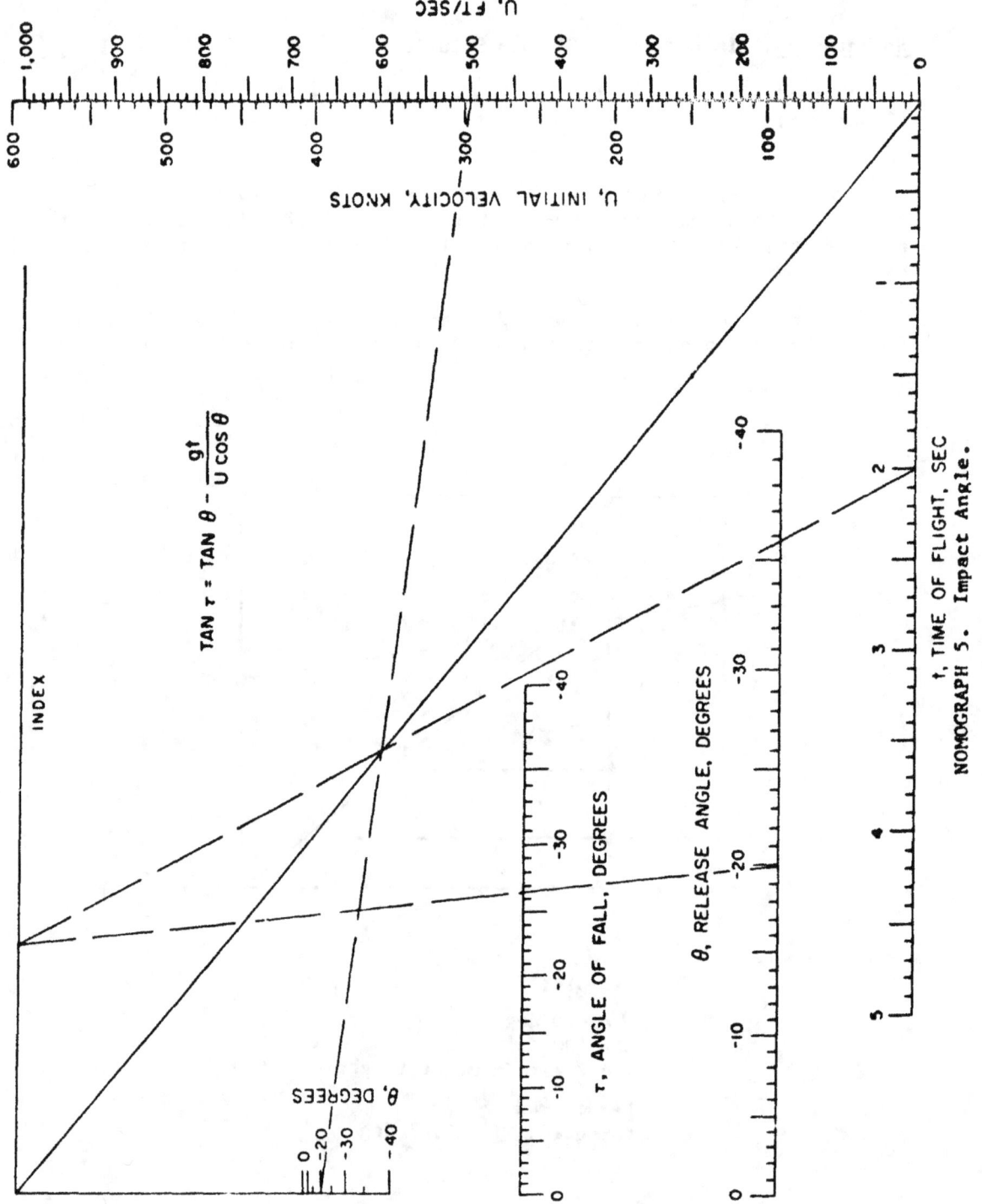

NOMOGRAPH 5. Impact Angle.

Nomograph 6. Altitude (Vacuum Solutions)

This nomograph solves the relation

$$Z = Ut \sin \theta + \frac{1}{2} gt^2 ,$$

Given any three of the four variables Z, U, t, or θ, the fourth may be determined. Use of the nomograph is straightforward.

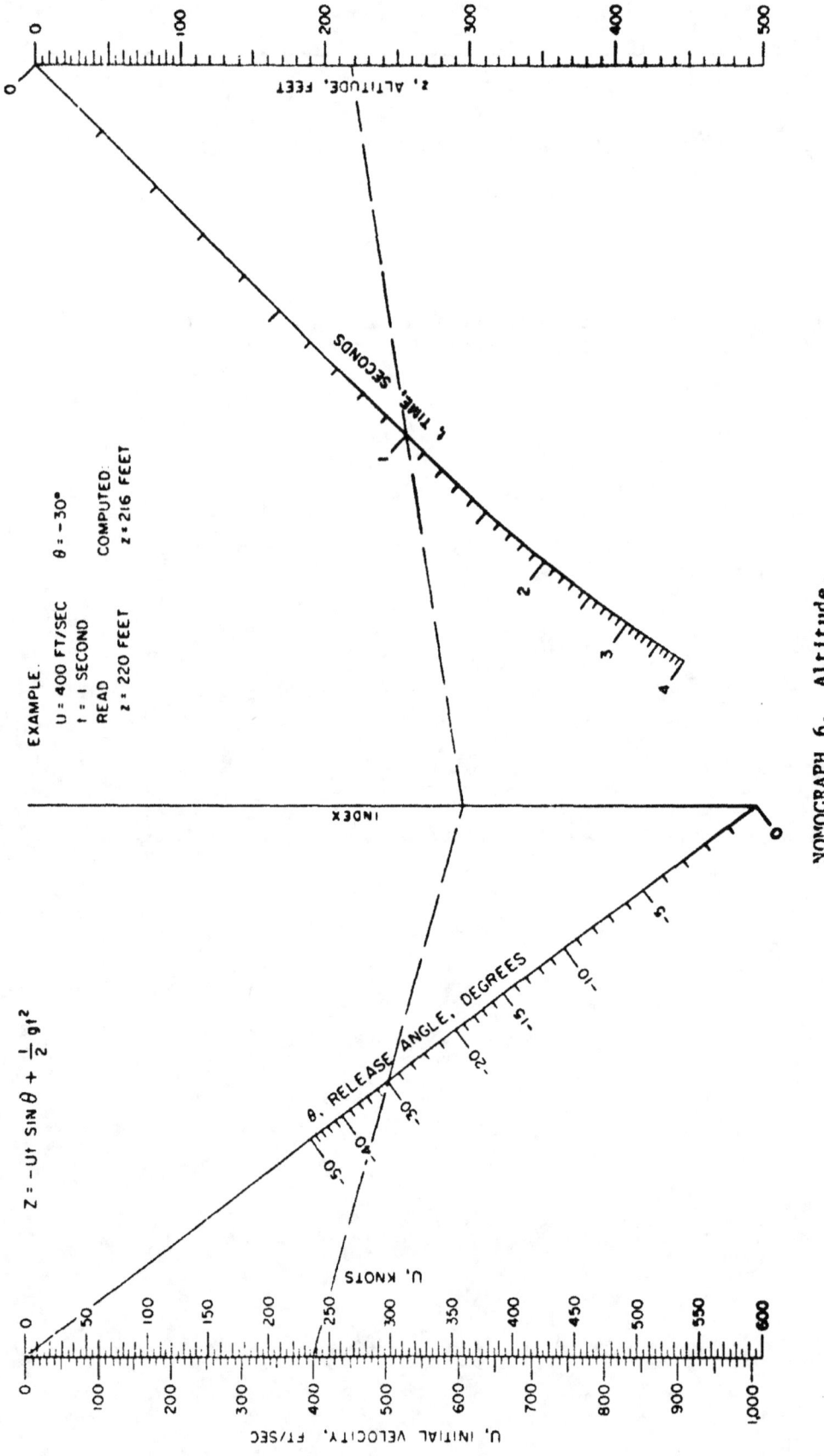

NOMOGRAPH 6. Altitude.

C. STANDARD DRAG BOMBS

Nomograph 7. Mach Number, K_D, cK_D Product, Altitude Corrections

This nomograph uses empirical data to determine several different quantities. Values of c, the reciprocal ballistic coefficient, are indicated for various bombs on one scale. Reference to the bomb data section (III.F.) will show that many bombs follow the K_D curves that are graphed here. If a bomb follows a K_D curve not graphed, its K_D values may be taken from the K_D curve section and this value may then be located on the far right scale.

Using the first three steps of the following procedure, the Mach number for a given velocity and altitude may be found. It is also possible to calculate another parameter, $1000 \, (\rho/\rho_0) \, cK_D$, that is used sometimes as a correction rather than merely taking the value of $1000 \, cK_D$.

It should be noted that with this nomograph, one need not necessarily start with step one of the procedure. Depending on the amount of information known beforehand, several steps may be eliminated. The complete procedure follows:

1. Locate release velocity, label as A.
2. Locate altitude on oblique scale, label as B.
3. Construct line AB, extend to Mach number scale. This is value of <u>Mach number</u> under given conditions; label as C.
4. From point C, go vertically to intersection with appropriate K_D curve, then horizontally right to K_D scale. This gives the value of the <u>ballistic drag coefficient</u> under the given conditions. If K_D is known from some other source, it may be located immediately without going through the preceding steps; label as point D.
5. Locate value of c, label as E.
6. Construct line DE, determine intersection of DE with $1000 \, cK_D$ scale. Note that units here are ft^2/lb; however, in most of the following work, units will be $ft^2/1000 \, lb$. Thus, the number read here gives the value of cK_D product in units of $ft^2/1000 \, lb$; label as F.
7. Locate altitude on far right scale, label as G.
8. Construct line GF and extend to far left, read off value of $1000 \, \rho/\rho_0 \, cK_D$.

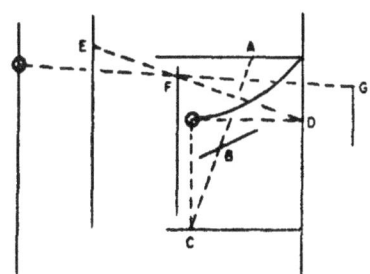

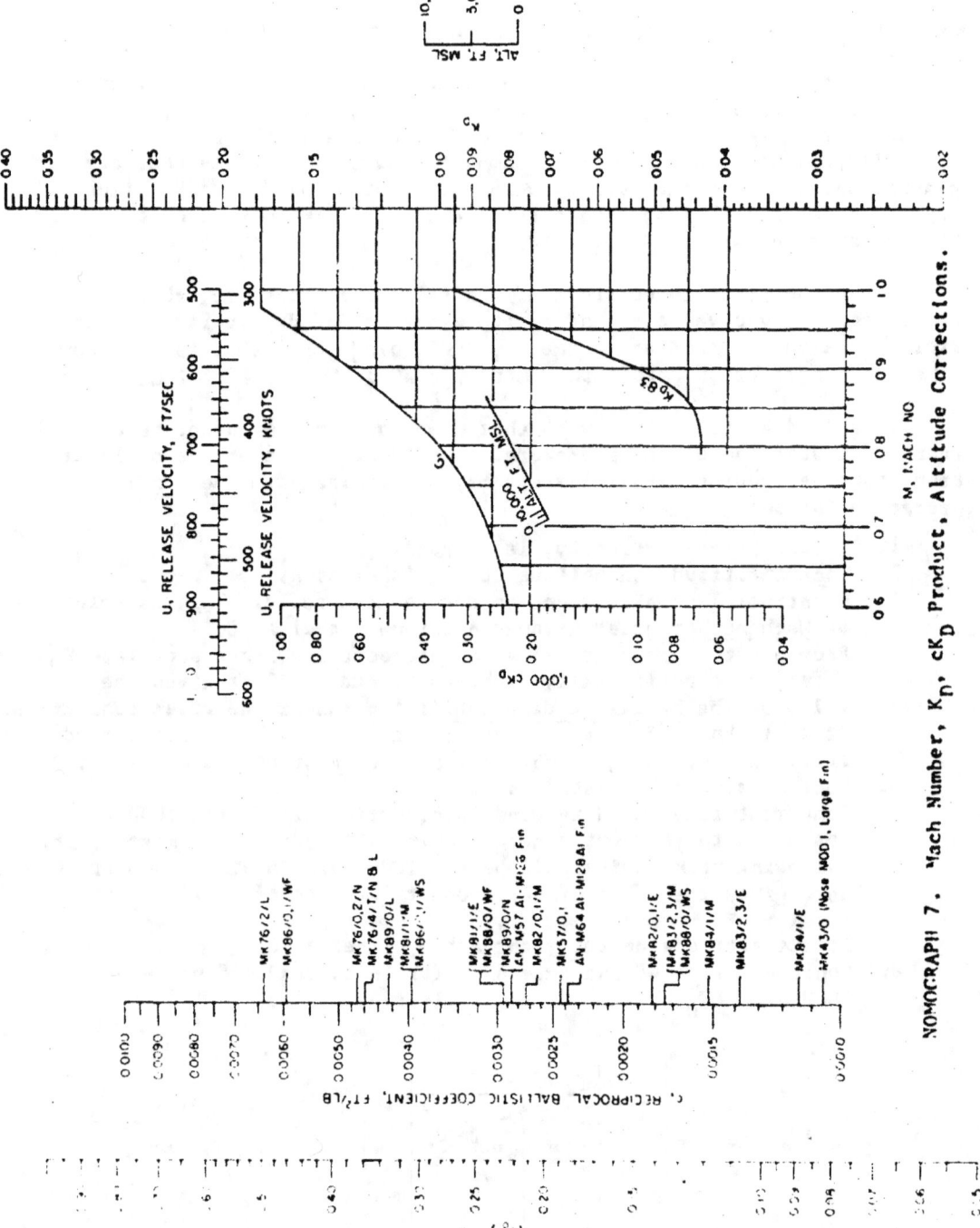

NOMOGRAPH 7. Mach Number, K_n, cK_D Product, Altitude Corrections.

NOTS TP 3902

Nomograph 8. Impact Angle (Standard Drag Bombs)

This nomograph uses eq. 51b to determine the impact angle.

Use of nomograph:

1. Use Nomograph 7 to determine the cK_D value. Locate this value on graph scale, label as A.
2. Locate altitude on left scale, label as point B.
3. Construct line AB, label intersection with oblique index line as point C.
4. Locate ground range on bottom left scale, label as D.
5. Construct line CD, label intersection of CD with horizontal index line as E.
6. Draw vertical line through E to intersection with appropriate θ curve, then proceed horizontally to right to vertical index line. Label point of intersection with index as F.
7. Locate velocity on far right scale, label as G.
8. Construct line FG, label intersection of FG with upper oblique index line as H.
9. Locate ground range on upper right scale, label as J.
10. Construct line HJ, label intersection with horizontal index as K.
11. Locate release angle on lower right scale, label as L.
12. Construct line KL. Read τ where KL crosses scale.

Example:
Mk 83/2&3/E Bomb
Z = 5000 ft
θ = -10 deg
U = 1000 fps
X = 12,620
Read: τ = 31.5 deg
Computed:
    τ = 32.16 deg.

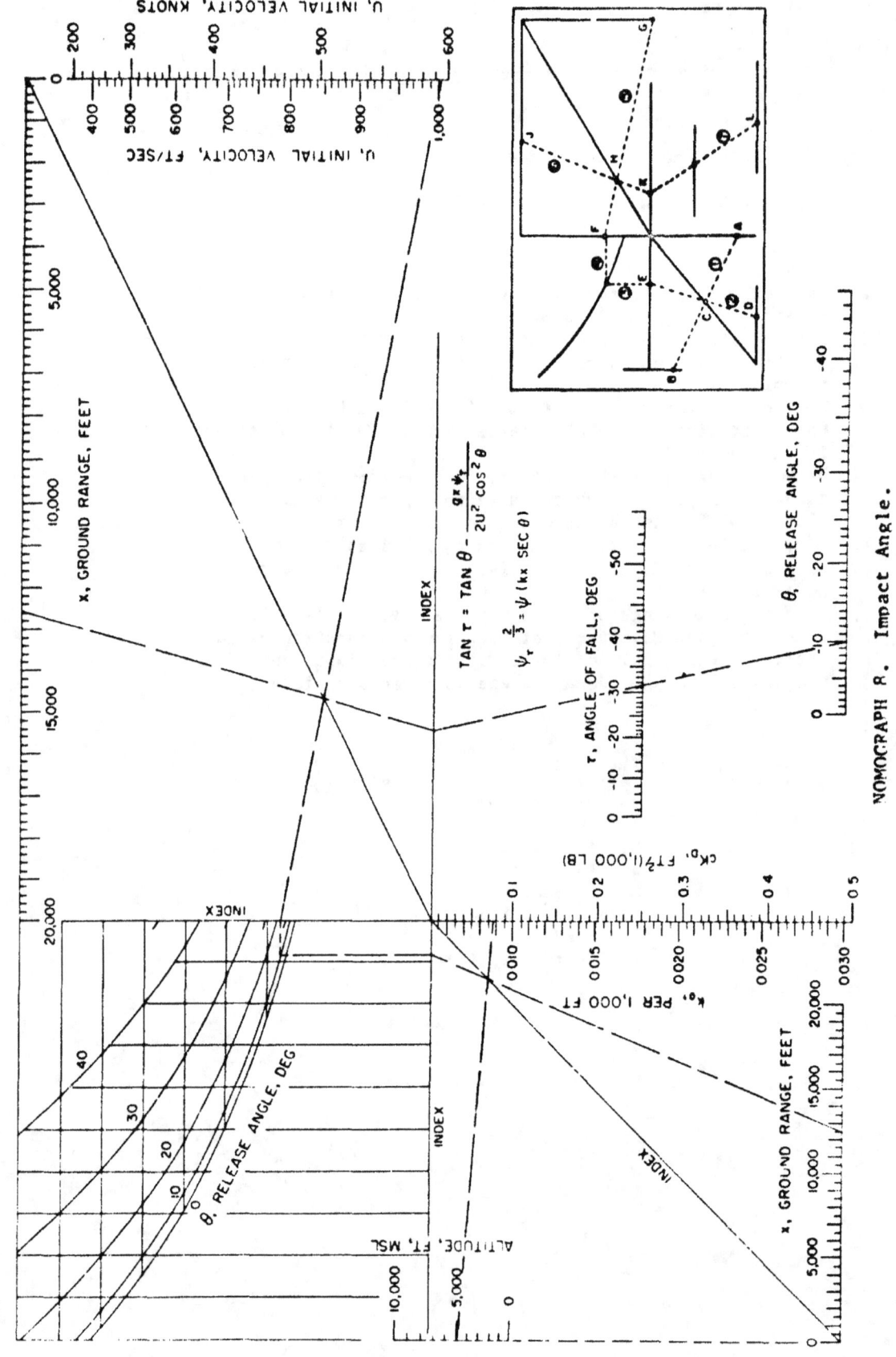

NOMOGRAPH R. Impact Angle.

Nomograph 9. Time of Flight or Ground Range (Standard Drag Bombs)

This nomograph uses a slightly modified form of eq. 52; here "a" is chosen for best fit rather than using 3/4.

Given X, t may be determined; and given t, X may be found. Instructions are for getting X, given t. The other method should be obvious once this case is studied.

Use of nomograph:

1. Locate time of flight on left scale, label as A.
2. Locate velocity U, label as B.
3. Construct line AB, label as C intersection of AB and left index line.
4. Determine value of cK_D from Nomograph 7, locate value and label as D.
5. Locate altitude on oblique scale, label as E.
6. Construct line DE, label as F intersection of DE with right index.
7. Construct line CF, label as G intersection of CF and curved scale.
8. Connect G to index point and extend line to left index, label as H.
9. Locate angle θ, label as J.
10. Construct line HJ, extend HJ to X-scale, read off X.

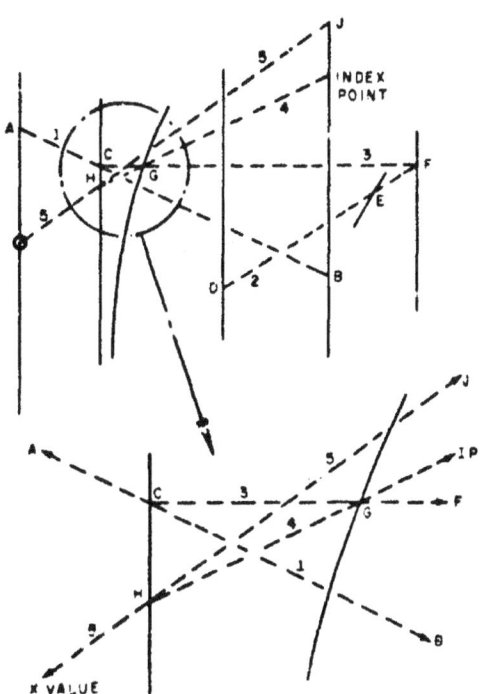

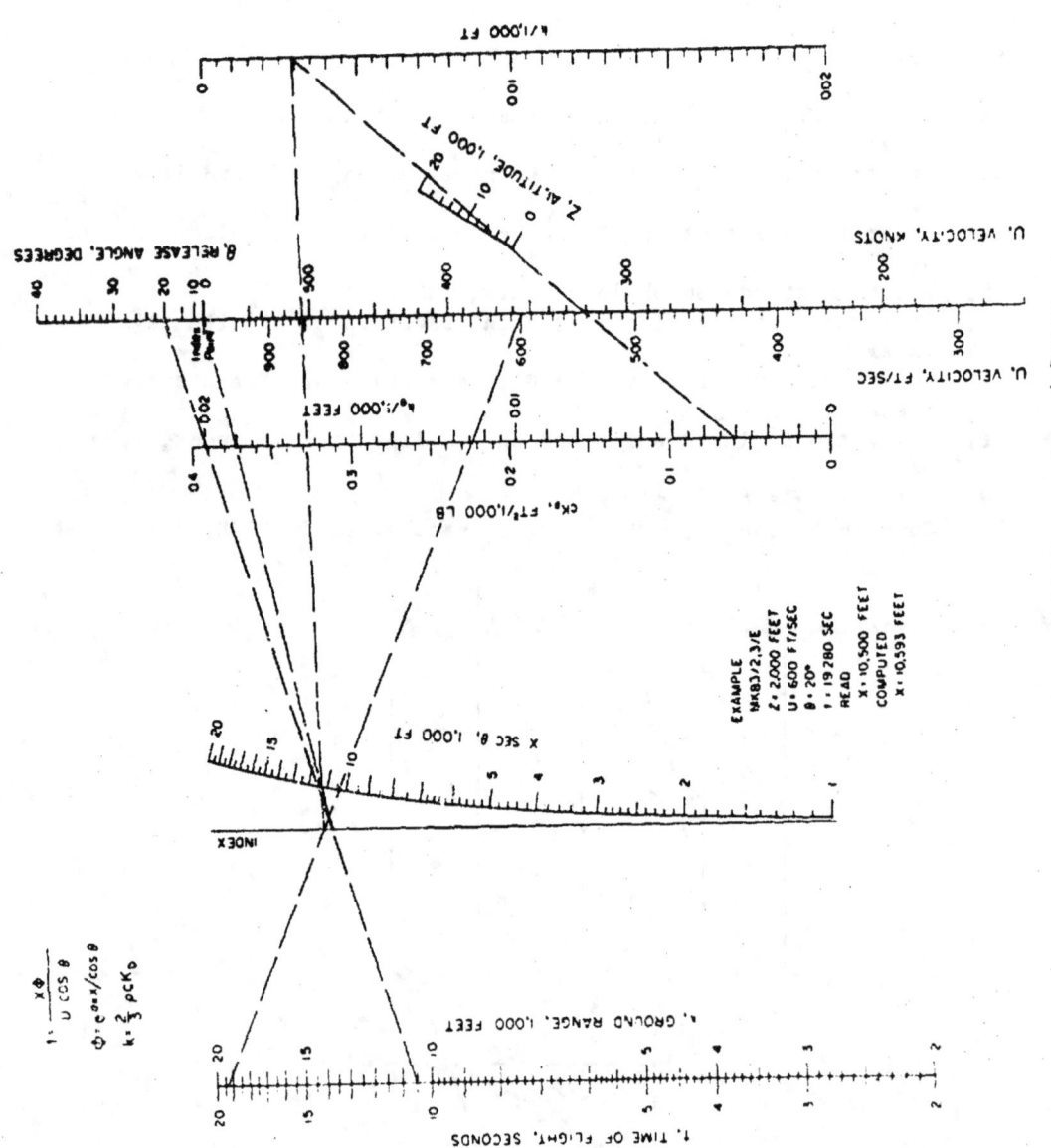

NOMOGRAPH 9. Time of Flight, or Ground Range.

Nomograph 10. Altitude (Standard Drag Bombs)

Using eq. 51a, this nomograph allows calculation of altitude Z for standard drag weapons.

By estimating altitude, cK_D may be determined from Nomograph 7. If original estimate is too inaccurate, the process may be repeated after answer is read from this nomograph, i.e., use iteration scheme for altitude.

Use of nomograph:

1. Locate ground range on upper left scale, label as A.
2. Locate value of cK_D, label as B on upper oblique scale.
3. Construct line AB, label as C intersection of AB and left vertical index.
4. From point C, proceed horizontally to appropriate θ curve, then down to horizontal index line. Label point of intersection on horizontal index line as D.
5. Locate velocity U on lower scale, label as E.
6. Construct line DE, label as F the intersection of DE and oblique index line.
7. Locate ground range on lower left vertical scale, label as G.
8. Construct line FG, extend line to Z, with $\theta = 0$ scale, label as point H. If $\theta = 0$ deg, point H is the altitude.
9. If $\theta \neq 0$ deg, locate X on far right scale. If $\theta > 0$, use upper portion of X-scale. If $\theta < 0$, use lower portion of X-scale. Label as J.
10. Locate θ on appropriate oblique scale, label as K.
11. Construct line JK, label as L intersection of JK and right vertical index.
12. Construct line HL, intersection of HL with Z, altitude scale gives value of Z.

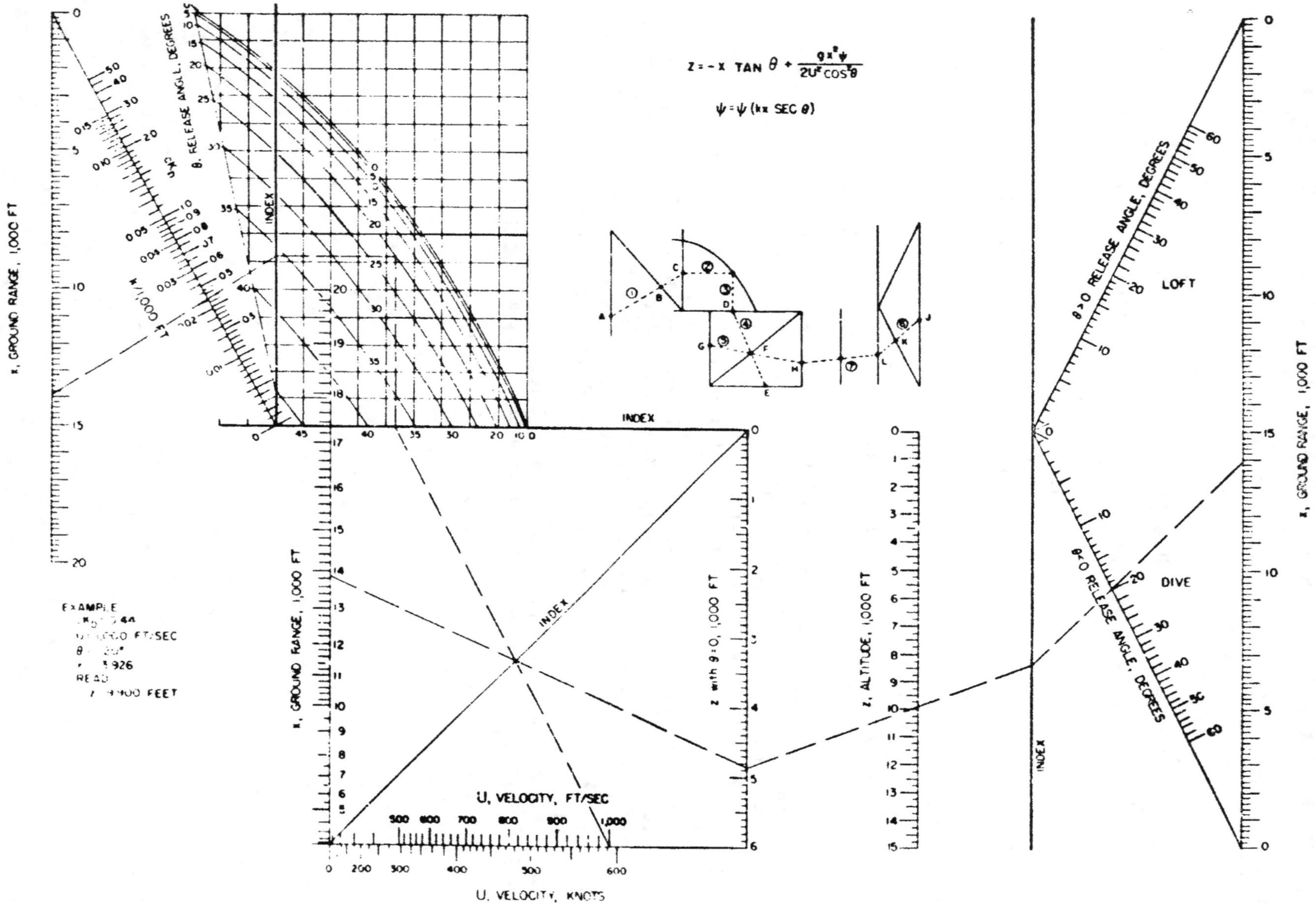

NOMOGRAPH 10. Altitude.

NOTS TP 3902

Nomograph 11. Altitude or Ground Range, Level Release (Standard Drag Bombs)

From this nomograph, ground range may be determined directly given the release altitude, velocity, and cK_D product of the weapon, or release altitude may be determined by iteration given ground range, velocity and cK_D product. The equation solved is

$$Z = \frac{gX^2}{2U^2} e^{\left(\frac{1}{\rho_0}\right) k_0 X} \; ; \quad k_0 = \left(\frac{2}{3}\right) \rho_0 cK_D .$$

Use: 1. Use Nomograph 7 to determine cK_D.

2. To find X, enter nomograph as shown in diagram on nomograph.

3. To find Z, estimate Z in step 2 of diagram.

4. Proceed in the step sequence 2, 3, 1 to find Z on the left-hand scale. If this Z does not agree with the initial estimate, enter step 2 with Z value just determined and continue as above until Z values agree.

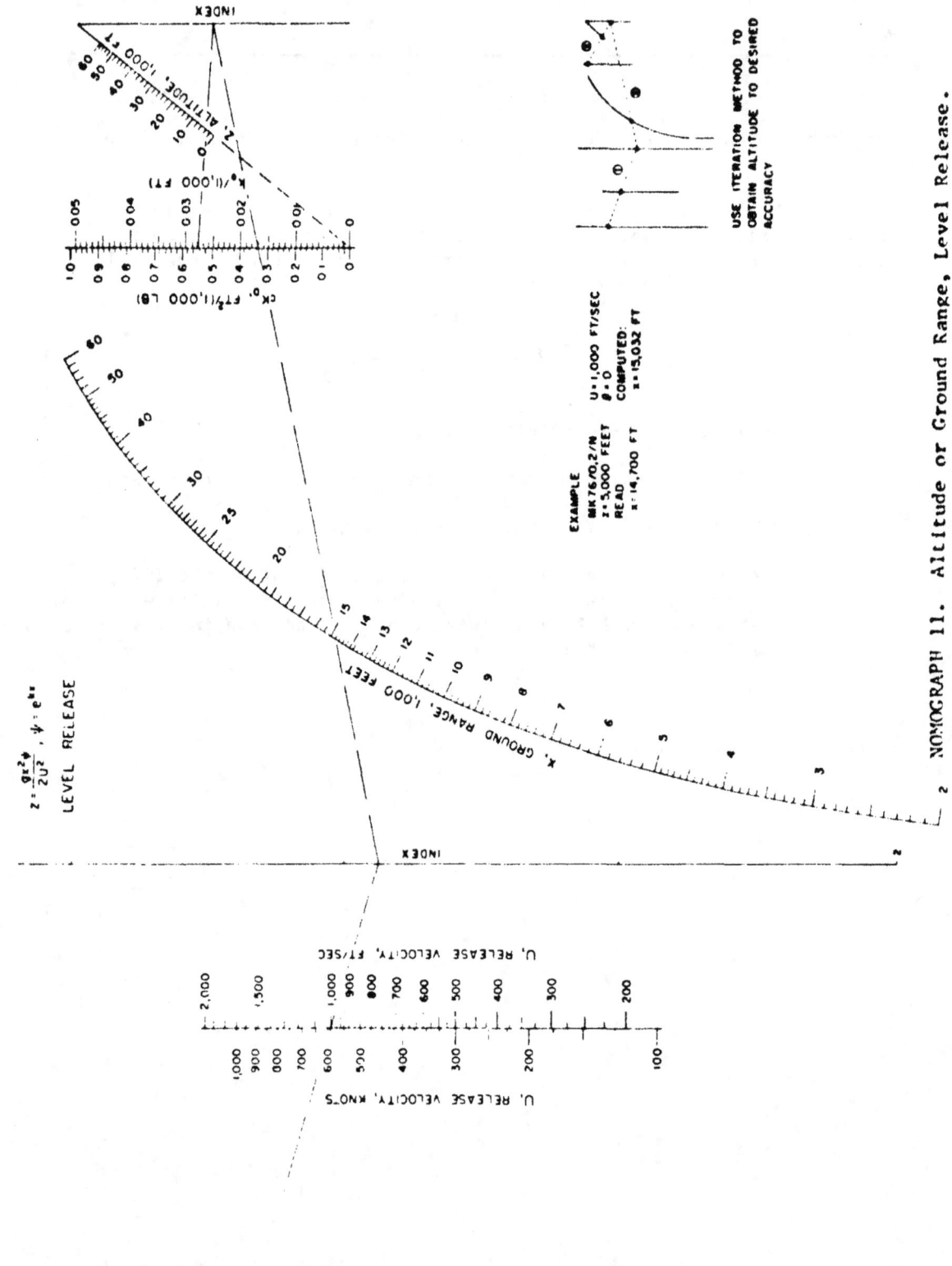

NOMOGRAPH 11. Altitude or Ground Range, Level Release.

Nomograph 12. Ground Range, Loft Bombing From $Z = 0$ to $Z = 0$. (Standard Drag Bombs)

This nomograph solves the equation

$$\sin \theta = \frac{gX\psi}{2U^2 \cos \theta}, \quad \psi = e^{kX}$$

for trajectories between $Z = U$ and $Z = D$.

Given any three of the four variables X, θ, U, or cK_D, the fourth can be found. The nomograph use for any of these cases is straightforward.

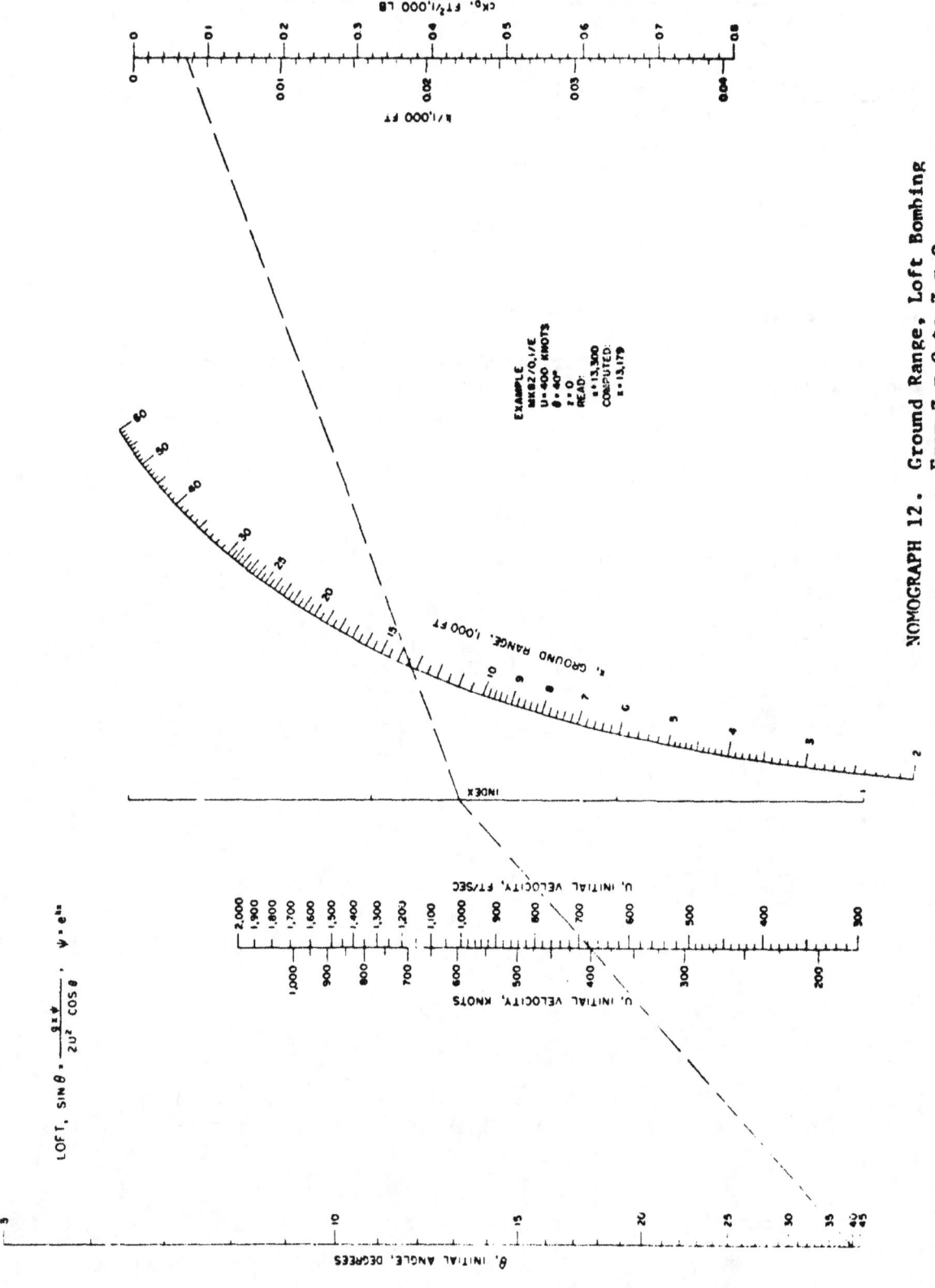

NOMOGRAPH 12. Ground Range, Loft Bombing From Z = 0 to Z = 0.

NOTS TP 3902

Nomograph 13. Ballistic Lead Angle Using Ground Range Data (Standard Drag Bombs)

This nomograph uses eq. 59 to solve for the ballistic lead angle directly. It can easily be used to solve for ground range or release angle using some iteration.

Use of nomograph:

1. Locate ground range on small horizontal scale to left, label as point A.
2. Locate altitude on scale, label as point B.
3. Construct line AB, extend AB to intersection with horizontal index line, label intersection as C.
4. Locate θ on scale in lower left corner, label as D.
5. Construct line CD, label as E intersection of CD with oblique index line.
6. Using Nomograph 7, locate and label as F the value of cK_D.
7. Construct line EF, label as G intersection of EF with vertical X-scale.
8. Locate velocity on right scale, label as H.
9. Construct line GH, label as J intersection of GH with center index line.
10. Locate X on left vertical scale, label as K.
11. Construct line JK, label intersection of JK with U-vertical scale as L.
12. Locate θ on vertical scale, label as M.
13. Construct line LM, intersection of LM with γ-scale gives the value of γ.

For cK_D information, refer to nomograph 7.

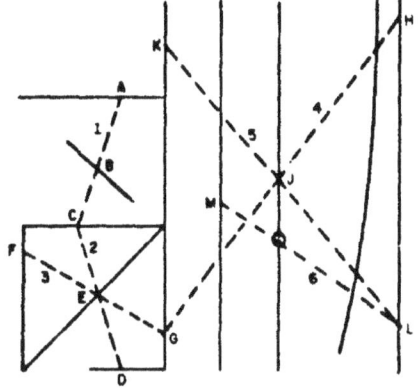

112

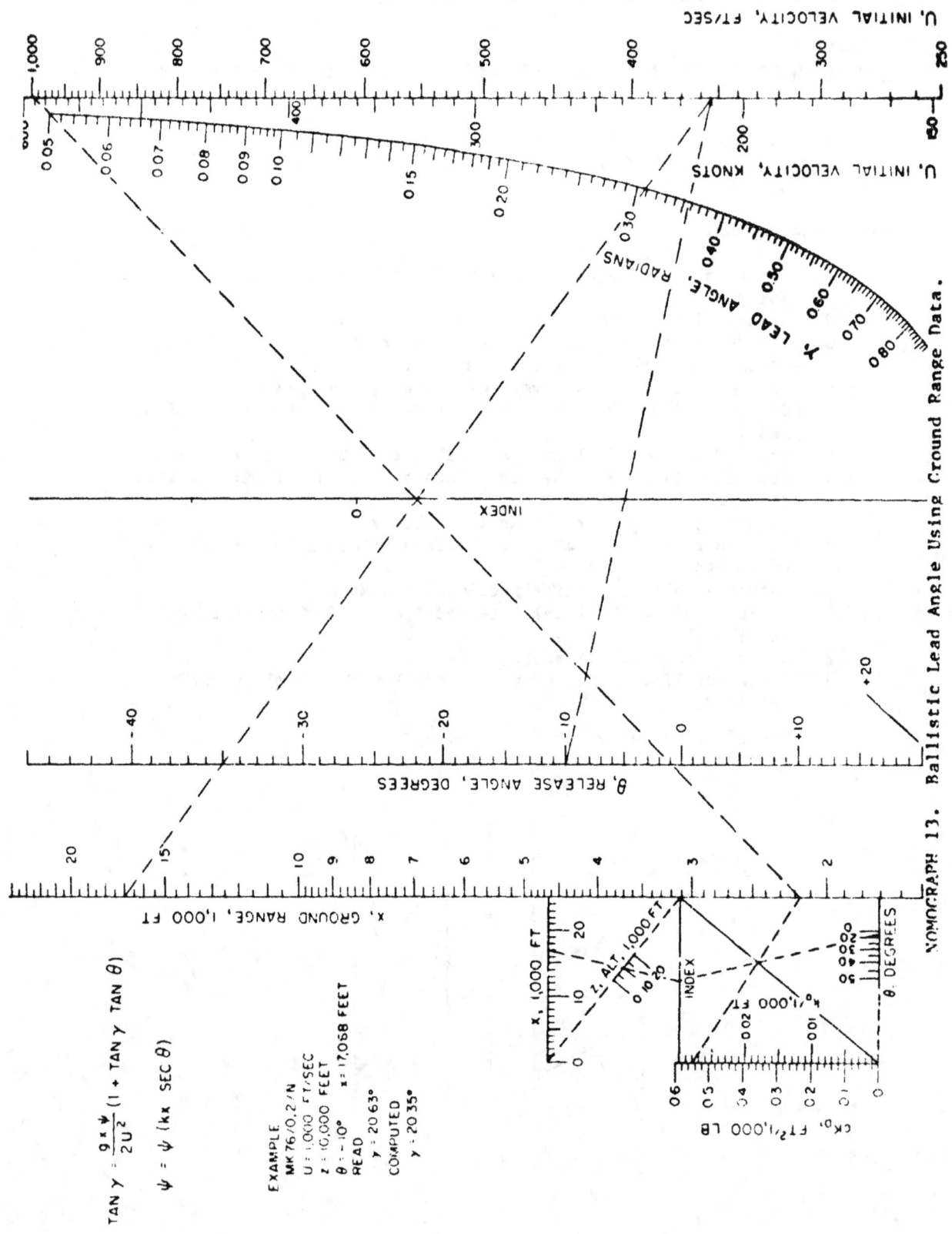

NOMOGRAPH 13. Ballistic Lead Angle Using Ground Range Data.

NOTS TP 3902

Nomograph 14. Ballistic Lead Angle Using Altitude Data (Standard Drag Bombs)

This nomograph allows calculation of the ballistic lead angle from altitude and airspeed information. The nomograph uses eq. 59.

Nomograph 7 may be used to determine the cK_D value for the various bombs. Necessary bomb data can be obtained from the K_D curve (III.C.) and Bomb Data (III.F.) sections.

Use of nomograph:

1. Locate velocity on left scale, label as A.
2. Locate altitude on center vertical scale, label as B.
3. Construct line AB, label as C intersection of AB and vertical index line.
4. Locate altitude on upper right scale, label as D.
5. Locate velocity on lower right scale, label as E.
6. Construct line DE, label as F the intersection of DE and oblique index line.
7. Locate and label as G the value of cK_D.
8. Construct line FG, label as H the intersection of FG and right index line.
9. Construct line CH, intersection of CH with correct release angle curve gives value of γ.

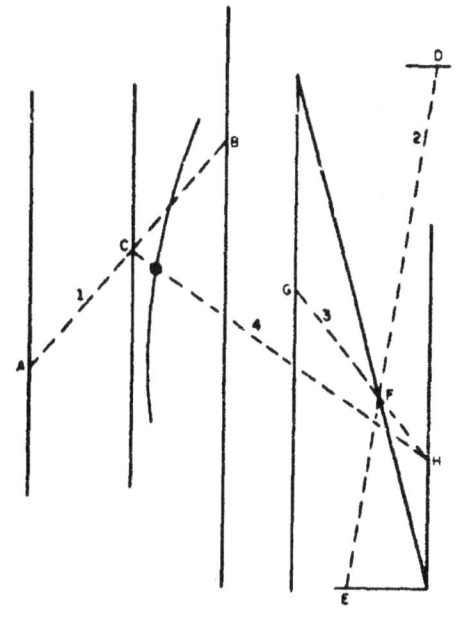

114

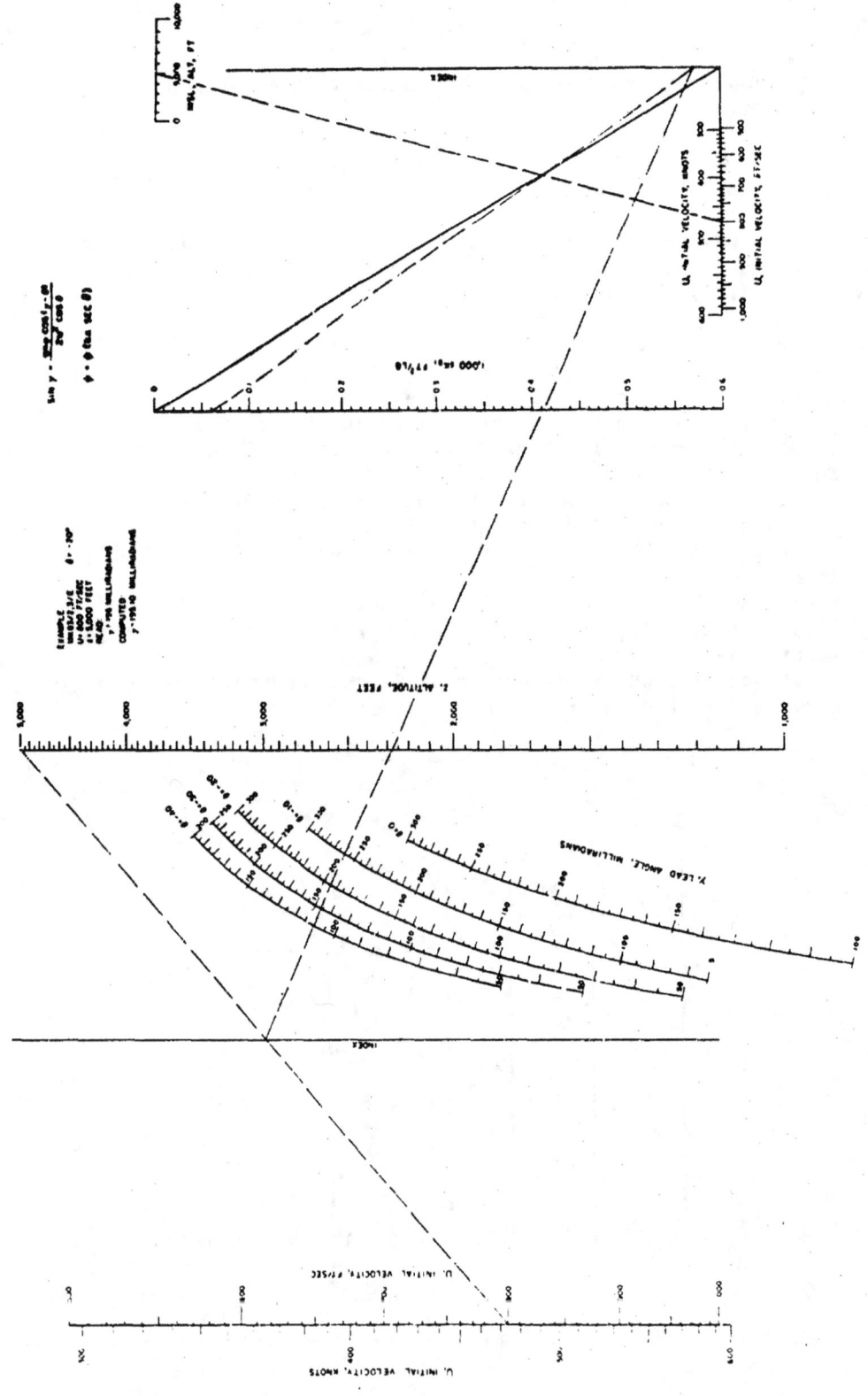

NOMOGRAPH 14. Ballistic Lead Angle Using Altitude Data.

Nomograph 15. Change in Ground Range Due to Small Change in Angle of Release About Level Release. (Standard Drag Bombs)

This nomograph solves the relation

$$\frac{\Delta X}{\Delta \theta} = \frac{U^2}{g} e^{-\frac{3}{2} kX} \quad ; \quad \theta = 0, \ Z = \text{constant},$$

giving the ratio of the change in ground range, ΔX, due to a small deviation of release angle, $\Delta \theta$, about level release.

Use:
1. Determine cK_D product from Nomograph 7.
2. With known X, Z, U, and the cK_D product, the sequential steps are indicated in the chart diagram.

Note that the ratio $\Delta X/\Delta \theta$ is given in ft/m rad. Multiply by 17.34 to obtain the ratio in ft/degree.

$$\frac{\Delta x}{\Delta \theta} = \frac{U^2}{g} e^{-\frac{3}{2}kx}, \quad \theta = 0, \quad z \text{ CONSTANT}$$

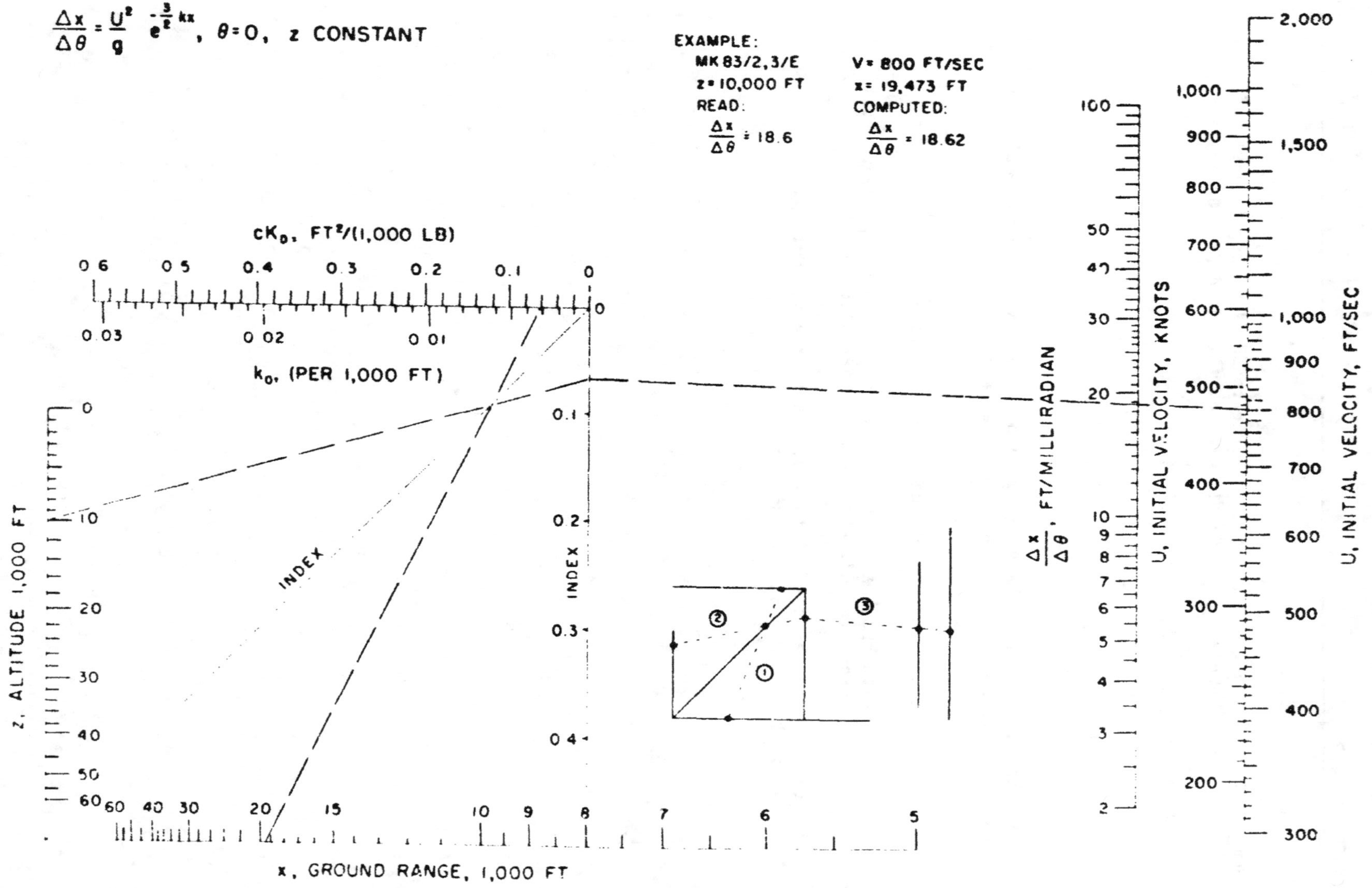

NOMOGRAPH 15. Change in Ground Range Due to Small Change in Angle of Release About Level Release.

D. RETARDED BOMBS

Nomograph 16. Mach Number, K_D, cK_D Product, Altitude Corrections (Retarded Bombs)

This nomograph uses empirical data to allow the determination of several different quantities. Reference to the Bomb Data (III.F.) section will show that many bombs follow the K_D curves that are graphed here. If a bomb follows a K_D curve not graphed, its K_D values may be taken from the K_D curve section and this value may then be located on the far right scale.

Using the first three steps of the following procedure, the Mach number for a given velocity and altitude may be found. It is possible to calculate another parameter, $1000 (\rho/\rho_0) cK_D$, which is used sometimes as a correction rather than merely taking the value of $1000 cK_D$.

It should be noted that with this nomograph, one need not necessarily start with step one of the procedure. Depending on the amount of information known beforehand, several steps may be eliminated. Complete procedure follows:

1. Locate release velocity, label as A.
2. Locate altitude on oblique scale, label as B.
3. Construct line AB, extend to Mach number scale. This is value of Mach number under given conditions; label as C.
4. From point C, go vertically to intersection with appropriate K_D curve, then horizontally to right to K_D scale. This gives the value of the ballistic drag coefficient under given conditions. If K_D is known from some other source, it may be located immediately without going through the preceding steps; label as D.
5. Locate value of c using Bomb Data (III.F.) section, label as E.
6. Construct line DE, determine intersection of DE with $1000 cK_D$ scale. Note that units here are ft^2/lb; however, in most of following work, units will be $ft^2/1000$ lb. Thus the number read off here gives the value of cK_D product in units of $ft^2/1000$ lb. Label as F.
7. Locate altitude on far right scale, label as G.
8. Construct line FG and extend to far left, read off value of $1000 \rho/\rho_0 cK_D$.

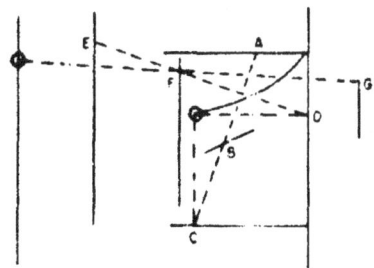

118

NOTS TP 3902

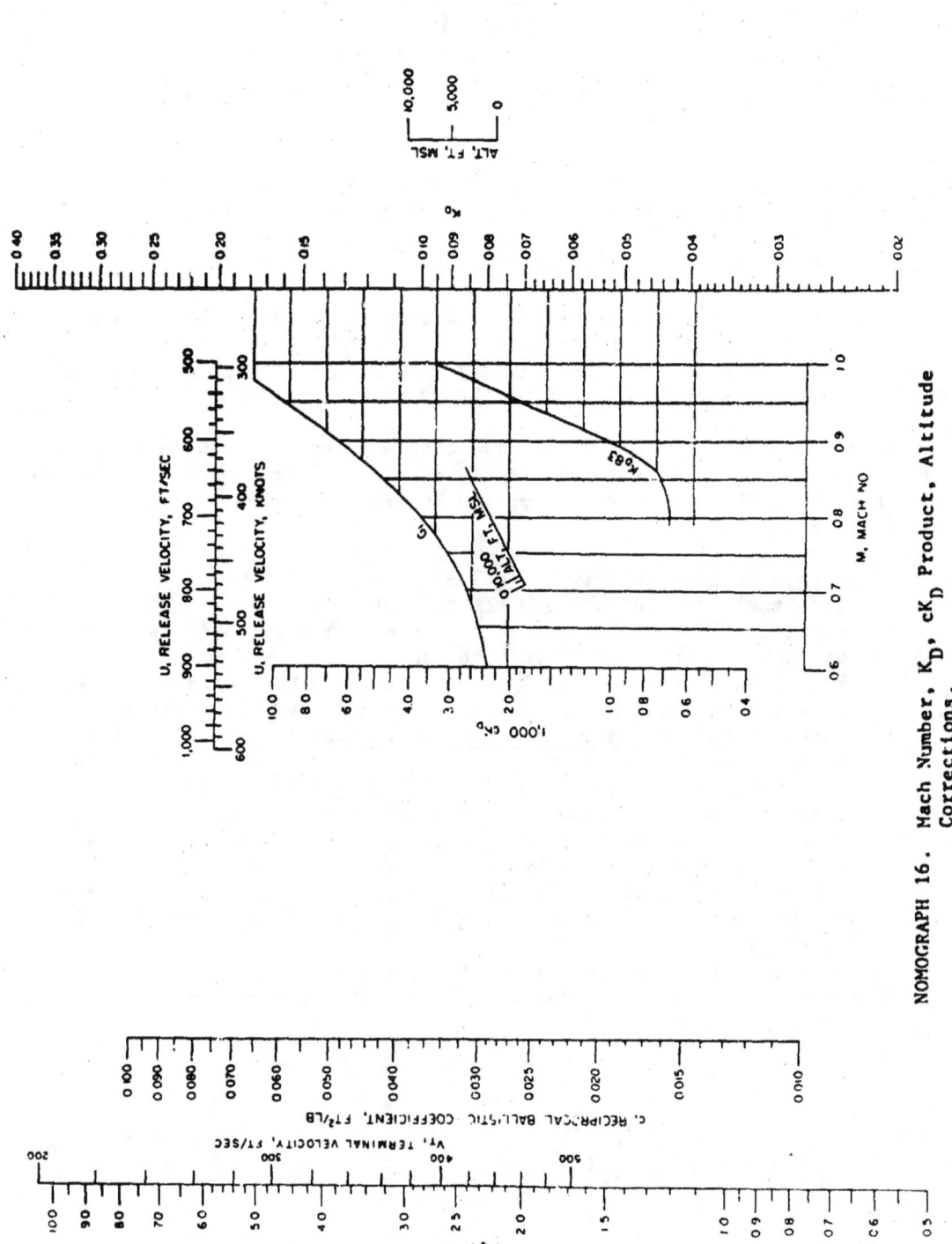

NOMOGRAPH 16. Mach Number, K_D, cK_D Product, Altitude Corrections.

119

NOTS TP 3992

Nomograph 17. Impact Angle (Retarded Bombs)

This nomograph uses eq. 51c to allow the determination of impact angles for retarded bombs.

Use of nomograph:

1. Locate ground range on upper left scale, label as A.
2. Locate terminal velocity of K, label as B.
3. Construct line AB, label as C intersection of AB and left vertical index.
4. From point C, proceed horizontally toward the right until appropriate θ curve is intersected. Then proceed down to horizontal index, label as D.
5. Locate velocity, label as E.
6. Construct line DE, label as F intersection of DE and oblique index line.
7. Locate ground range on lower left vertical scale, label as G.
8. Construct line FG, label as H intersection of "τ at θ = 0 deg" scale. If θ = 0 deg, read impact angle at H.
9. If $\theta \neq 0$ deg, locate θ on far right scale, label as J.
10. Construct line HJ, read τ where HJ crosses scale.

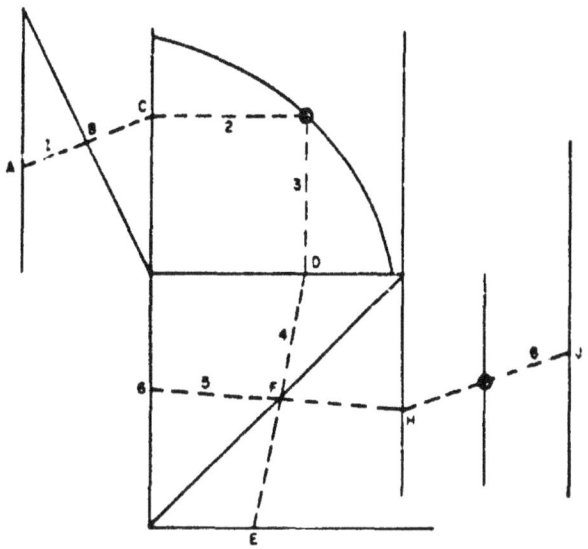

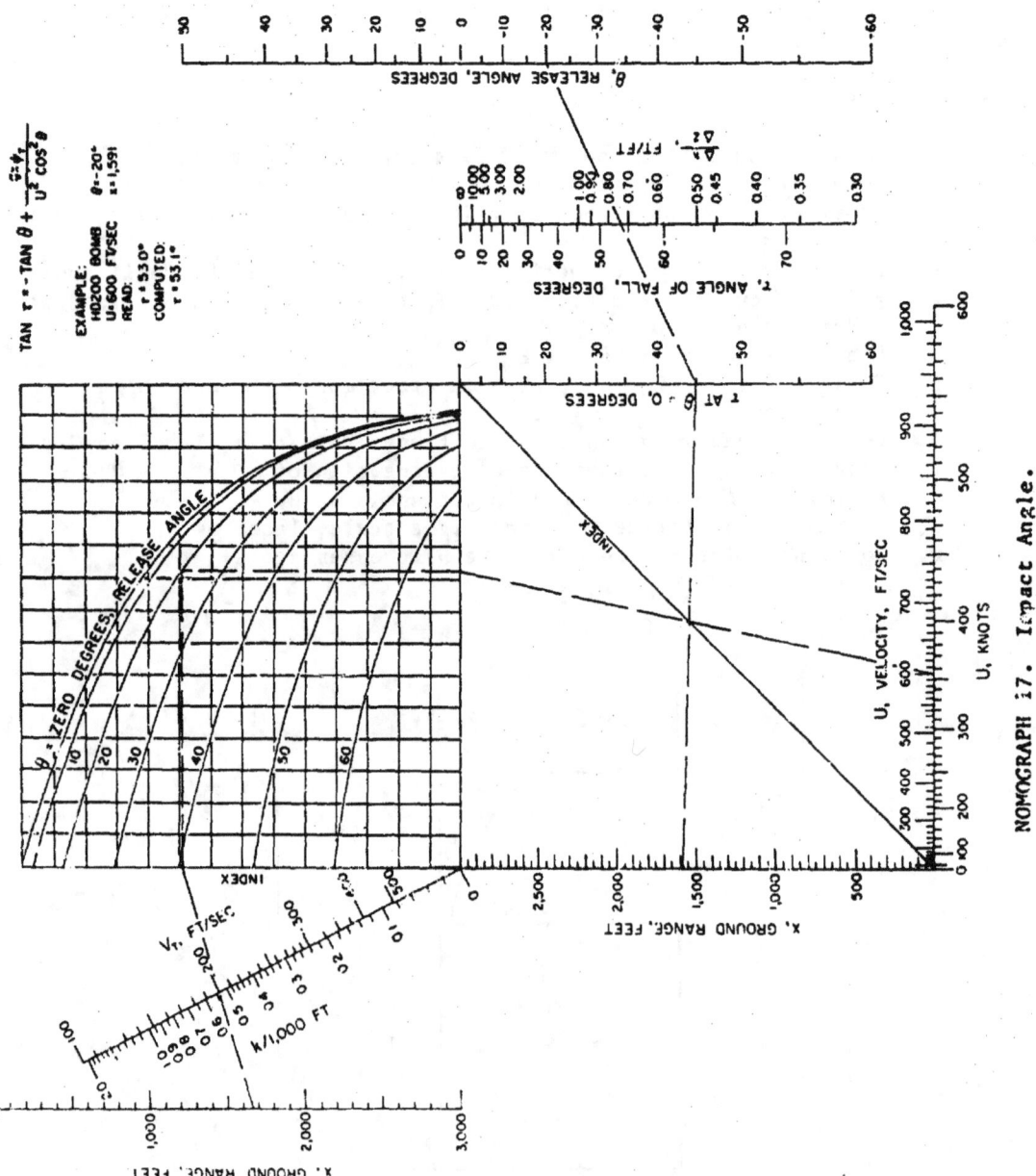

NOMOGRAPH 17. Impact Angle.

Nomograph 18. Time of Flight or Ground Range (Retarded Bombs)

This nomograph uses eq. 52.

Use of nomograph:

1. Locate ground range on left scale, label as A.
2. Locate θ on oblique scale, label as B.
3. Construct line AB; label as C intersection of AB with vertical "v_t, terminal velocity" scale.
4. Locate initial velocity on right scale and label as D.
5. Construct line CD; label as E intersection of CD with oblique index line.
6. From point C, proceed horizontally until appropriate terminal velocity curve is intersected, then proceed vertically to upper horizontal scale; label point on horizontal scale as F.
7. Construct line EF; time of flight given by intersection of EF with lower horizontal scale.

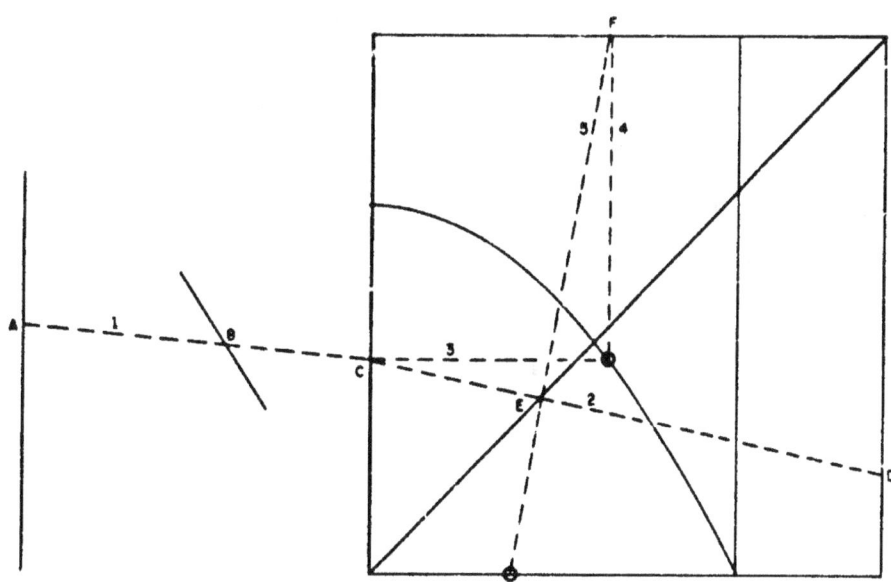

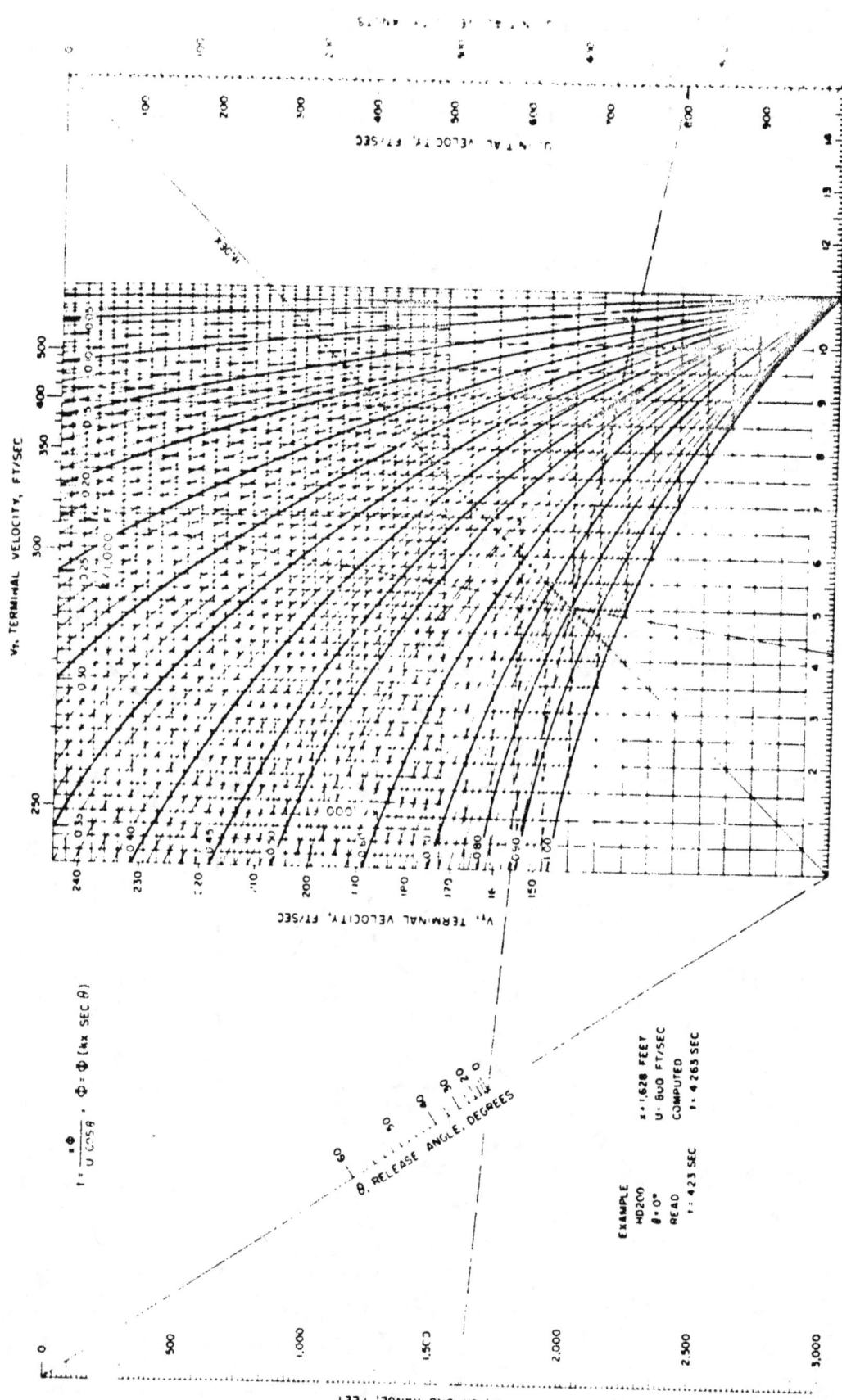

NOMOGRAPH 18. Time of Flight of Ground Range.

Nomograph 19. Altitude (Retarded Bombs)

This nomograph uses eq. 51a. It should be noted that a small change in ground range can cause a large change in altitude. This can be verified by reference to the bomb tables, and may be understood by recalling that for the retarded weapons, the X-component of the velocity is damped out rapidly. Thus the velocity vector of the bomb is directed almost perpendicular to the earth after a certain flight time.

Use of nomograph:

1. Locate desired ground range on upper left scale, label as A.
2. Locate terminal velocity on upper oblique scale, label as B.
3. Construct line AB, label as C intersection of AB with left index line.
4. From point C, follow horizontal over to appropriate θ curve, then down a vertical line until horizontal index line is intersected. Label intersection as D.
5. Locate initial velocity on lower scale, label as E.
6. Construct line DE, label as F intersection of DE and oblique index line.
7. Locate desired ground range on lower left scale, label as G.
8. Construct line FG, label intersection of FG and "Z, with θ = 0 deg" scale as H. If θ = 0 deg, point H gives value of Z.
9. If $\theta \neq 0$ deg, locate correct value of θ on two oblique scales to right of graph. Label as point J.
10. Locate ground range on appropriate portion of extreme right scale. If $\theta > 0$, use upper portion; if $\theta < 0$, use lower portion. Label as K.
11. Construct line JK, label as L intersection of JK and right index line.
12. Construct line HL, read altitude where HL crosses scale.

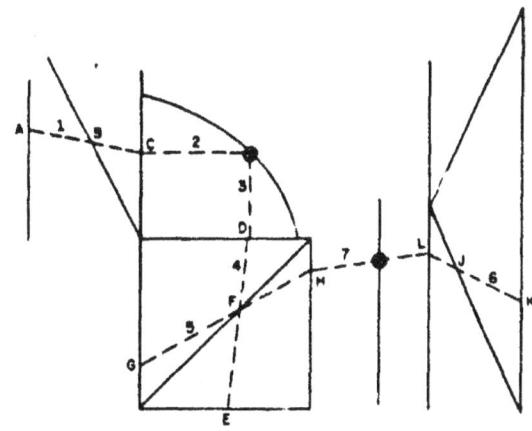

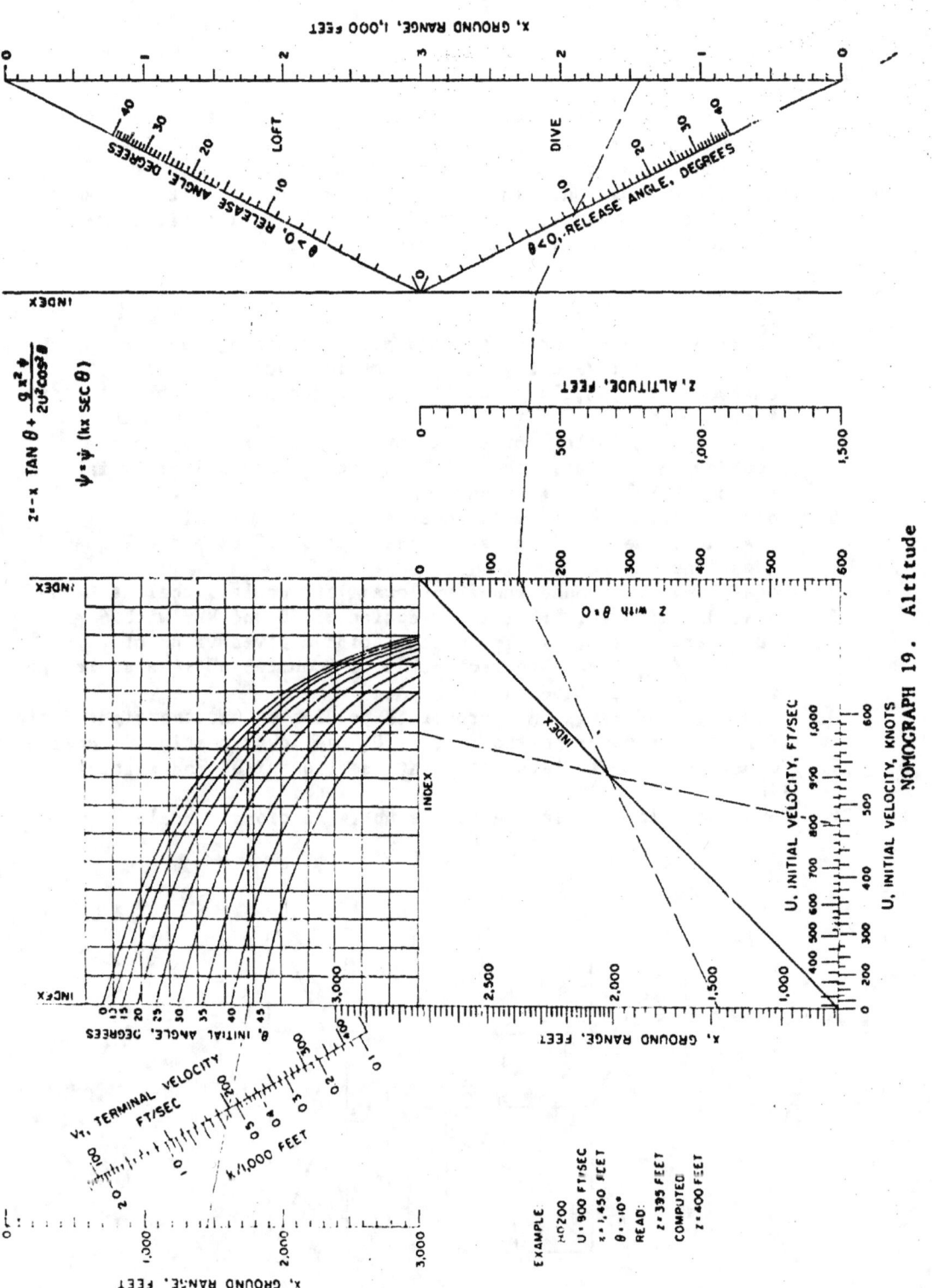

NOMOGRAPH 19. Altitude

NOTS TP 3902

Nomograph 20. Ground Lag, Ground Range, or Altitude, Level Release
(Retarded Bombs)

This nomograph uses a variation of eq. 51a for level release conditions. With this nomograph, it is possible to determine the ground lag, ground range, and given altitude. The nomograph may also be used to calculate altitude, given ground range, but this suffers from the instability pointed out in nomograph 19.

Use of nomograph:

1. Locate altitude, label as A.
2. Locate velocity, label as B.
3. Construct line AB, label as C the intersection of AB and index line.
4. Locate terminal velocity (or k*), label as D. (k* is chosen to give a best fit in writing $\psi = \exp k^* X \sec \theta$).
5. Construct line CD; ground range is given by intersection of CD with X-line.
6. Construct line from point C to point k* = 0 deg (or infinite terminal velocity). Note intersection of this line with X-line. This number given bomb range in a vacuum.
7. Ground lag is given by difference of answers in steps 5 and 6.

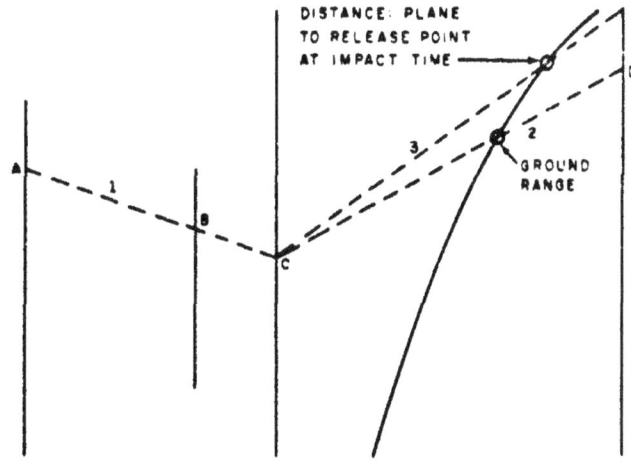

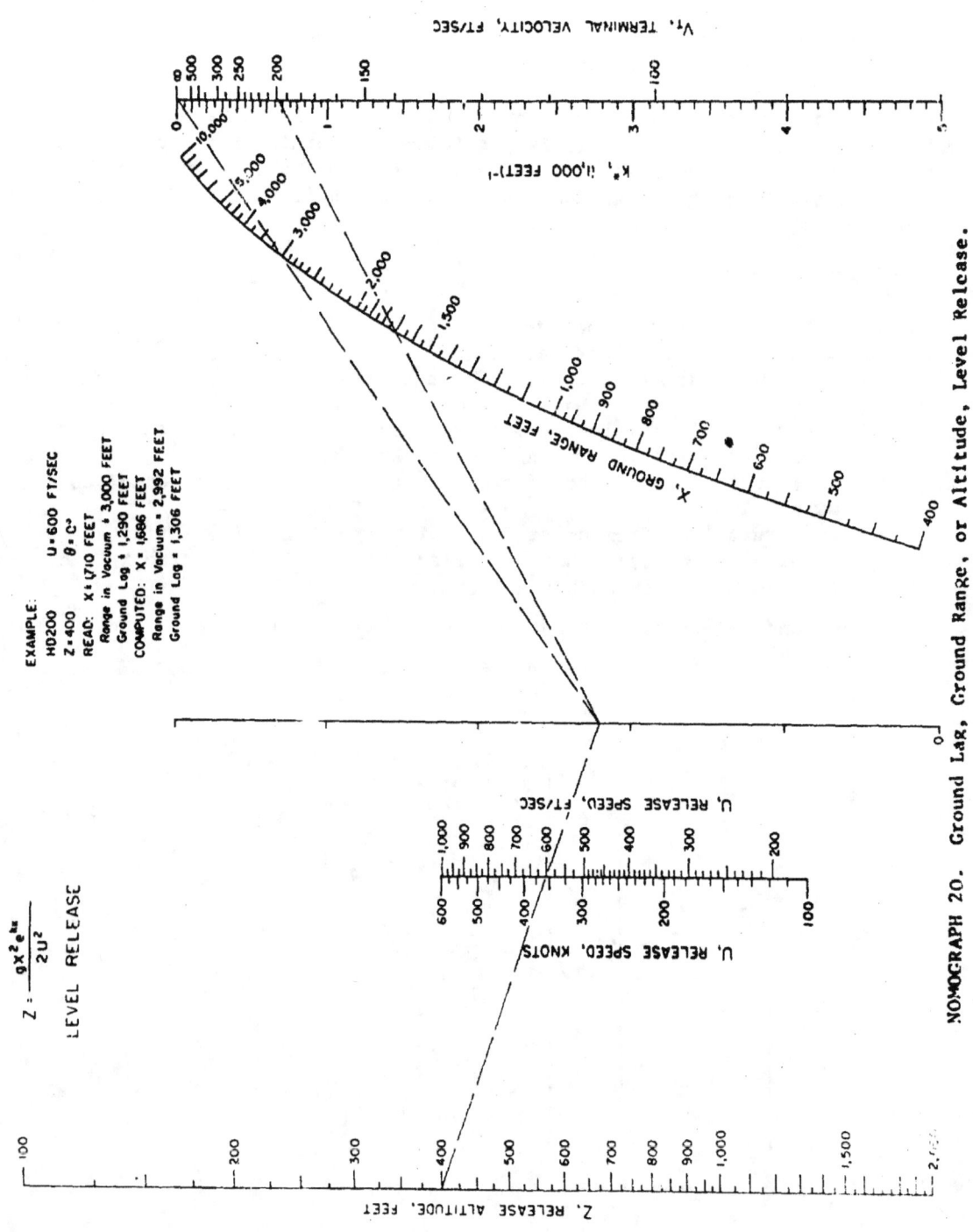

NOMOGRAPH 20. Ground Lag, Ground Range, or Altitude, Level Release.

NOTS TP 3002

Nomograph 21. Ballistic Lead Angle Using Ground Range Data (Retarded Bombs)

This nomograph uses eq. 54b.

Use of nomograph:

1. Locate terminal velocity on upper horizontal scale, label as A.
2. Locate θ on lower scale, label as B.
3. Locate ground range on left vertical scale, label as C.
4. Construct line AB, label as D intersection of AB and index line.
5. Construct line CD, label as E intersection of CD with center index line.
6. Note that right side of center index line is divided into groups of ten subdivisions, as is left side of center "X-feet" scale. If point E is a certain number of subdivisions below a main dividing line, locate the point on the left side of X-feet scale which is an equal number of subdivisions below the same main dividing line. Label as F.
7. Locate U on far right scale, label as G.
8. Construct line FG, label as H intersection of FG and right index line.
9. Locate ground range on X-feet scale in center, label as J.
10. Construct line JH, label as K intersection of JH with U-scale.
11. Locate θ on center vertical scale, observe sign of θ, label as L.
12. Construct line KL, read γ where KL intersects γ-scale.

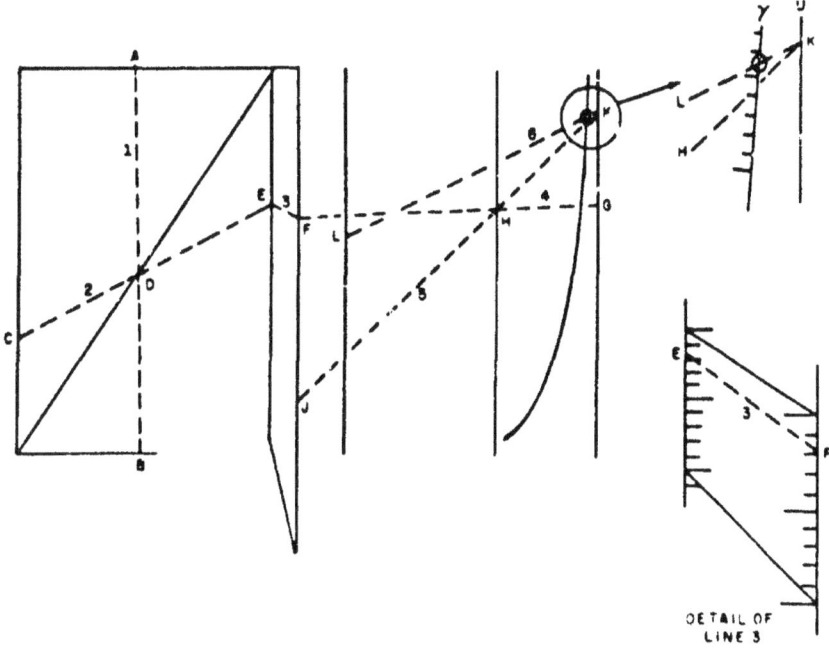

128

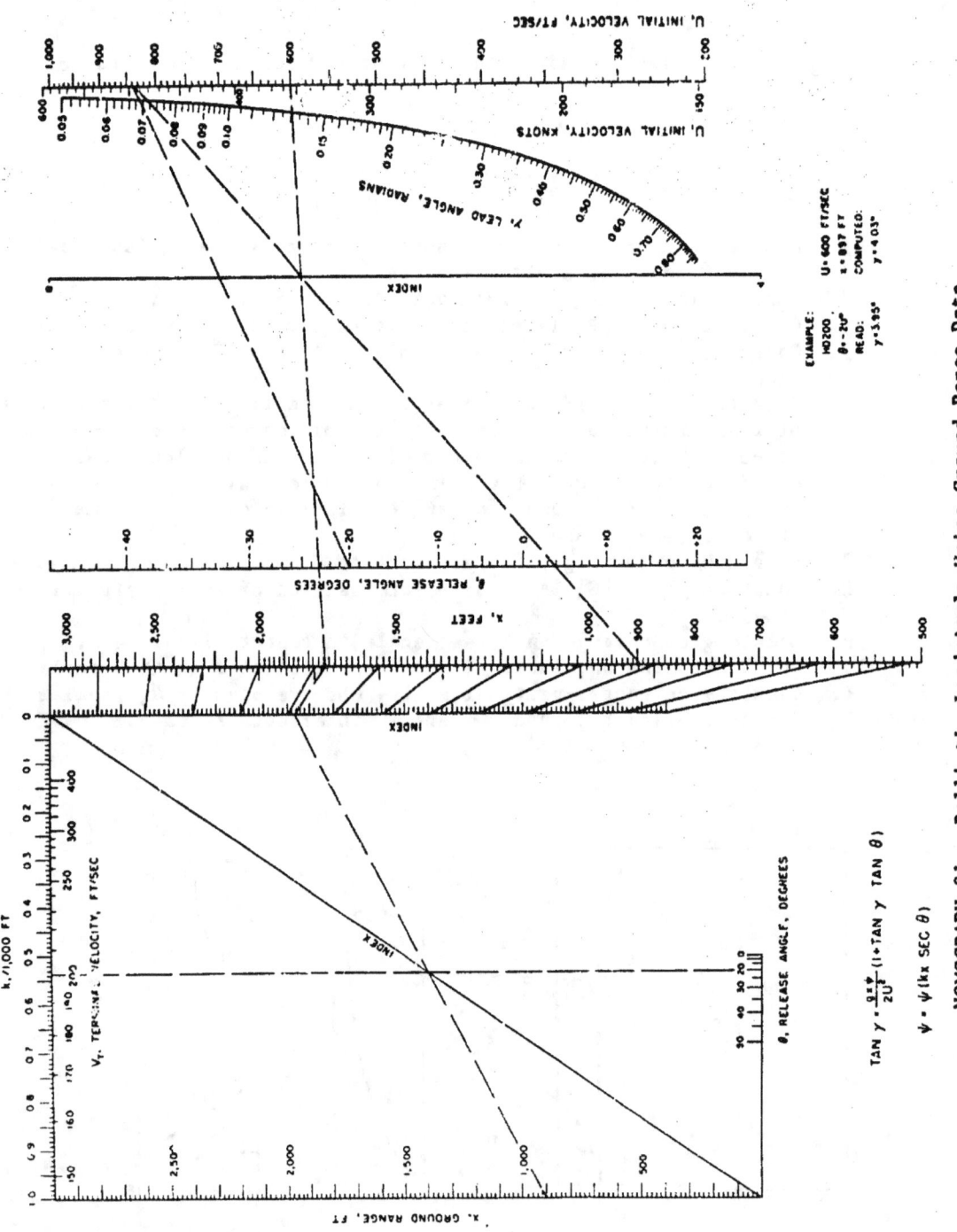

NOMOGRAPH 21. Ballistic Lead Angle Using Ground Range Data.

NOTS TP 3902

Nomograph 22. Change in Ground Range Due to Small Change in Angle of Release About Level Release. (Retarded Bombs)

This nomograph solves the relation

$$\frac{\Delta X}{\Delta \theta} = \frac{X^2}{2Z(1 + 1/2 \ln \psi)} \quad , \quad \ln \psi \doteq k^* X \; , \; \theta \doteq 0°$$

$$k^* = f(k) \quad , \quad Z = \text{constant}$$

giving the ratio of the change in ground range, ΔX, due to a small deviation of release angle, $\Delta \theta$, about level release.

Use:
1. The value of terminal velocity V_T, if not known, can be obtained from Nomograph 16.
2. With V_T or k^*, X, and Z known, the solution steps are as indicated by the diagram in the nomograph.

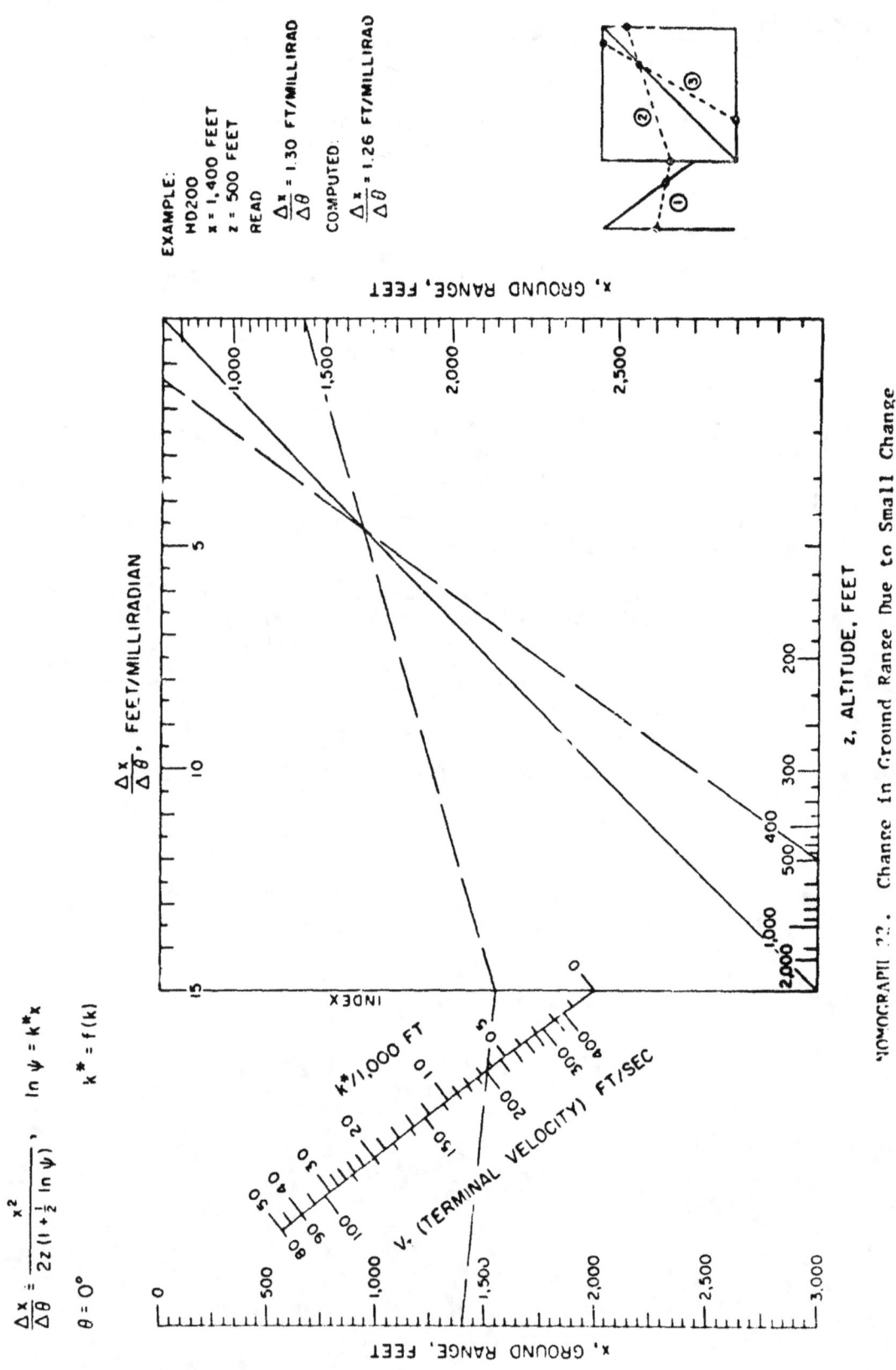

NOMOGRAPH 23. Change in Ground Range Due to Small Change in Angle of Release About Level Release.

NOTS TP 3902

<u>Nomograph 23.</u> Time of Flight or Altitude, Level Release. (Retarded Bombs)

This nomograph solves the relation

$$t = \sqrt{\frac{2Z}{g}}\ e^{k_{Zt} Z^{1/3.5}}$$

Given V_T and Z, t may be found, or, given V_T and t, Z may be found. V_T may be obtained from Nomograph 16.

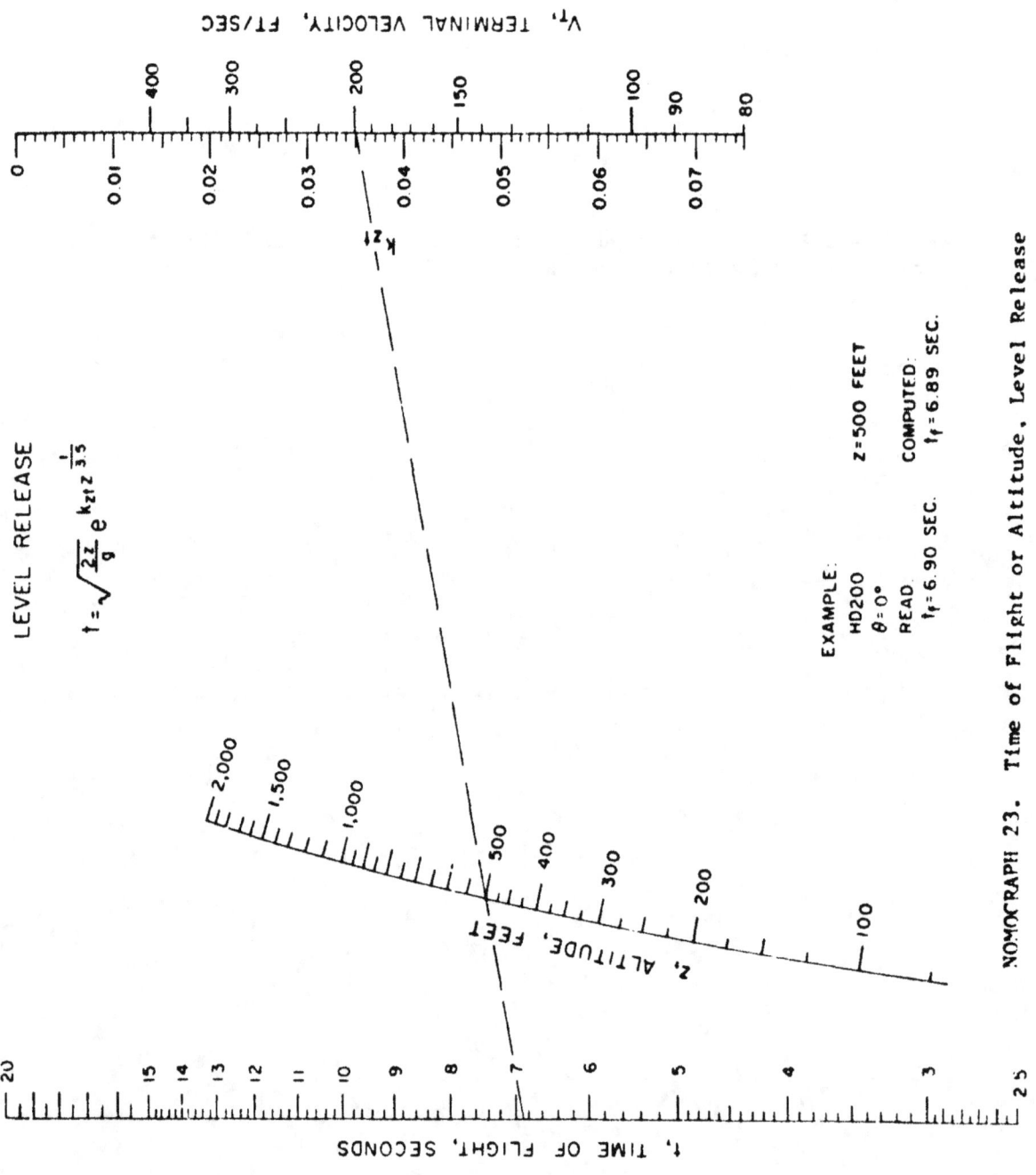

NOMOGRAPH 23. Time of Flight or Altitude, Level Release

NOTS TP 3902

INITIAL DISTRIBUTION

30 Chief, Bureau of Naval Weapons
    CD (1)   RAAV-33 (1)
    CW (1)   RAAV-51 (1)
    R (1)   RAAV-81 (1)
    DLI-31 (2)   RM (1)
    R-2 (1)   RM-3 (1)
    R-3 (1)   RM-371 (1)
    RA (1)   RM-373 (1)
    RA-2 (1)   RMGA-4 (1)
    RA-3 (1)   RMWC (1)
    RAAV (1)   RMWC-4 (1)
    RAAV-22 (1)   RT (1)
    RAAV-3 (2)   S (1)
    RAAV-30 (2)   SEL (1)
    RAAV-32 (1)
4 Chief of Naval Operations
    OP-506C (2)
    OP-722C1 (2)
1 Chief of Naval Research (Code 104)
1 Air Development Squadron 5
1 Fleet Air Whidbey
1 Heavy Attack Squadron 2 (VAH 2)
1 Heavy Attack Squadron 4 (VAH 4)
1 Heavy Attack Squadron 6 (VAH 6)
1 Heavy Attack Squadron 8 (VAH 8)
1 Heavy Attack Wing 1, Sanford
1 Naval Air Development Center, Johnsville
1 Naval Air Force, Atlantic Fleet
1 Naval Air Force, Pacific Fleet
1 Naval Air Test Center, Patuxent River
2 Naval Avionics Facility, Indianapolis (Technical Library)
1 Naval Missile Center, Point Mugu (Technical Library)
1 Naval Ordnance Laboratory, Corona
1 Naval Ordnance Laboratory, White Oak (Technical Library)
1 Naval Postgraduate School, Monterey
1 Naval Research Laboratory
2 Naval Weapons Laboratory, Dahlgren (Technical Library)
1 Navy Electronics Laboratory, San Diego
1 Nuclear Weapons Training Center, Atlantic
1 Operational Test and Evaluation Force
1 Navy Liaison Officer, Tactical Air Command, Langley Air Force Base
1 Aberdeen Proving Ground (Ballistic Research Laboratories)
1 Army Engineer Research and Development Laboratories, Fort Belvoir
   (STINFO Branch)
1 Picatinny Arsenal (Technical Library)
1 White Sands Missile Range (ORDBS-Technical Library)

1 Yuma Test Station
 1 Tactical Air Command, Langley Air Force Base (TPL-RQD-M)
 1 Air Proving Ground Center, Eglin Air Force Base
 1 Air University Library, Maxwell Air Force Base (AUL-8236)
 1 Tactical Air Warfare Center, Eglin Air Force Base (VC)
 1 Systems Engineering Group, Deputy for Systems Engineering, Wright-Patterson Air Force Base (SEPRR)
20 Defense Documentation Center (TISIA-1)

U. S. NAVAL ORDNANCE TEST STATION
CHINA LAKE, CALIFORNIA

IN REPLY REFER TO
751/CEV:ga
17 November 1965

From: Commander, U. S. Naval Ordnance Test Station
To: Distribution of NOTS Technical Publication 3902

Subj: NOTS TP 3902, <u>Ballistic Handbook</u>, dated July 1965; transmittal of errata sheet for

Encl: (1) Errata sheet dated 17 November 1965 for subject report

1. It is requested that the corrections described on the enclosed errata sheet be incorporated in NOTS TP 3902.

C. E. VAN HAGAN
By Direction

NOTS TP 3902

ERRATA

Title page: Change the publishing date of August 1965 to read July 1965

Abstract cards: Change the publishing date of August 1965 (3rd line) to read, July 1965.

17 November 1965
Enclosure (1)